T0350837

Mastering Organizational Change Management

Barbara A. Davis

J.ROSS
PUBLISHING

Copyright © 2017 by Barbara Davis

ISBN-13: 9781604271416

Printed and bound in the U.S.A. Printed on acid-free paper.

10 9 8 7 6 5 4 3 2 1

Library of Congress Cataloging-in-Publication Data
Names: Davis, Barbara, 1969– author.
Title: Mastering organizational change management / by Barbara A. Davis.
Description: Plantation, FL : J. Ross Publishing, [2017] | Includes
  bibliographical references and index.
Identifiers: LCCN 2017011755 | ISBN 9781604271416 (hardcover :   alk. paper)
Subjects: LCSH: Organizational change—Management.
Classification: LCC HD58.8 .D3747 2017 | DDC 658.4/06—dc23 LC record
available at https://lccn.loc.gov/2017011755

Phone: (954) 727-9333
Fax: (561) 892-0700
Web: www.jrosspub.com

Dedication

This book is dedicated to:

My husband Robert: for bringing out the best in me and inspiring me to reach higher every single day.

Andrea Montanti: for helping me to bring this and so many other literary projects to completion this past year.

Marty Clarke: for being an inspirational speaker and writer (and all-around great guy) and providing a much appreciated story for this material.

Contents

About the Author

Barbara A. Davis has been a champion for technology standards, infrastructure, and organizational change management for over 16 years. She is an international speaker who also works with Fortune 500 companies to realign business analysis services, critical and struggling projects, and establish operational infrastructure in order to ensure successful outcomes in the face of conflict and challenging circumstances.

Barbara came into technology with a degree in conflict resolution and over 15 years of functional business experience including professional training, project management, community development, business ownership, change management, and conflict resolution. She has drawn on these experiences throughout the course of her career and has become a change champion by defining the organizational capability through infrastructure (such as career paths, assessment tools, competencies, and key performance indicators), training (such as educational programs and workshops), and the creation of Centers of Excellence and management frameworks. She audits and redefines operational management of key practice areas and methodologies.

Throughout her career, Barbara has interviewed and assessed hundreds of resources, and has held titles and roles including Manager of Human Capital, Business and IT Portfolio Manager, IT Operational

Manager, Methodologist, Solutions Consultant, Project Manager, Business Analyst, Author, and Professional Skills Trainer. Her experiences include operational management, organizational change management, document management, vendor management, configuration management, change control, practice management, business analysis, project management, cyber security, and auditing PMO methodologies.

Barbara has published numerous articles and is author of the books *Managing Business Analysis Services: A Framework for Sustainable Projects and Corporate Strategy Success, Going Beyond the Waterfall: Managing Scope Across the Project Life Cycle,* and *Mastering Software Project Requirements.* She created the world's first university-accredited business analysis diploma program and has spoken at Project Summit/BA World Conferences across Canada, the United States, and India.

At J. Ross Publishing we are committed to providing today's professional with practical, hands-on tools that enhance the learning experience and give readers an opportunity to apply what they have learned. That is why we offer free ancillary materials available for download on this book and all participating Web Added Value™ publications. These online resources may include interactive versions of material that appears in the book or supplemental templates, worksheets, models, plans, case studies, proposals, spreadsheets and assessment tools, among other things. Whenever you see the WAV™ symbol in any of our publications, it means bonus materials accompany the book and are available from the Web Added Value Download Resource Center at www.jrosspub.com.

Downloads for *Mastering Organizational Change Management* include various impact assessment tools, a functional complexity matrix, and a change scorecard template to assist practitioners in managing a change project and measuring its success.

Introduction

MYTHS AND MISCONCEPTIONS ABOUT ORGANIZATIONAL CHANGE AND ORGANIZATIONAL CHANGE MANAGEMENT

Organizational change management (OCM) is increasingly critical as companies become more and more agile and socially oriented. Unfortunately, we are still holding onto past ideologies of what change management is and what it should be. These outdated ideologies guide and inform our attitudes and approaches to OCM.

Attitudes

Many organizations, and indeed many individuals, still hold the attitude that people (employees and customers alike) should just accept that changes are a part of life and readily adopt them. This attitude is a remnant of the archaic, oppressive management era that included the serfs and the *company man*. Monty Python had it precisely right when Dennis proclaimed that, "Supreme executive power derives from a mandate from the masses..."[1]

Thankfully, attitudes are changing. Perhaps we have the Millennials to thank for it. As a direct result of how this particular generation was raised, we are seeing a massive shift in the way companies operate and manage employees and customer relationships. It is out of this shift that we are also seeing a shift in emphasis on social currency. In other words, we are beginning to understand that people don't *have* to be loyal just because someone gave them a job. They don't have to be loyal because they can take what they learned from us over to our competitors or even start their own companies.

Ultimately, this means that existing companies must change the way they think and act on engagement to manage those critical relationships. All of this means that OCM must be front-of-mind whenever an organization is contemplating changes to the products or services they offer and even the ways in which they interact with employees and customers, partners, vendors, and yes, the competition.

Approaches

The most commonplace approach to OCM in the past at many companies was virtually nonexistent. Imagine trying to turn a car that is moving at full speed by suddenly yelling out *turn now* to the driver. Unfortunately, this is the type of approach that was taken in the past (and still is in many cases).

This means there was no planning, or visibility into the planning by people outside of the executive suite, no communication until the changes were to be implemented, no input taken into consideration, and little or no follow up. It was often referred to as a *top-down* approach to change management. This isn't an approach; it's a disaster.

IMPACT OF CHANGE ON CULTURE AND MORALE

Change can either have a positive or a negative impact on culture and morale. It all comes down to how the change is managed from start to finish, and how it is communicated and executed throughout that process.

The reality is that people want to feel as though they have some modicum of control over what happens to them and how they work. Naturally, change that is well managed is going to have a more positive impact on morale than change that is not.

INERTIA

Inertia is the ability of an object to continue on its trajectory and maintain its course without influence to slow it down or to stop it altogether. In grade school science, every child is taught that inertia is the ability for a body at rest to remain at rest and for a body in motion to stay in motion.

In fact, the same can be said for change. People and organizations can have a type of inertia to either continue along the same old bureaucratic path as they always have (hence: we've always done it this way) or to accept changes.

By employing the right techniques in the correct order, inertia can be used to help motivate people to change when the perception of success with those changes is high. Just as with a physical object, the momentum of the object can be redeployed and used to alter the course of that object. This is why the underlying barriers to change as well as the process of change are so critical for the practitioners to know and understand.

BUILDING A GRASSROOTS MOVEMENT

A grassroots movement is very much a *bottom-up* strategy in organizational change. Effectively, it is when the employees are engaged and motivated to embrace the impending changes and this, in turn, motivates the top levels of the organization to adopt the changes. There are good examples of this with social outcry in response to policies and actions. For example, people who are angered over the unjust firing of a good employee for giving a hungry child lunch in a school cafeteria. But there are also examples such as a company that dictates a new software must be used, but only a handful of people actually adopt it. The most powerful examples of grassroots movements come from political changes that we saw all around the world with the collapse of the Berlin Wall in 1989 and the March on Washington for Jobs and Freedom in 1963.

The grassroots movement is the polar opposite of the *top-down* approach, but can be successfully leveraged in parallel. In order to do this, there must be buy-in at both levels and activities that foster engagement and collaboration, while empowering people to participate.

CHANGE MANAGEMENT—BRACING FOR IMPACT

How will I do my job? Who will I report to? Will I lose my job? Why do we have to change? What's in it for me? How does this impact me personally?

We work so hard to fit into this world and find the niche we believe was created just for us. When something comes along to threaten that position—how we relate to the familiar environment and even who we interact with—it threatens our place in that niche.

The emerging field of change management is designed to address the specific issue of how major changes will impact the rank and file of corporations and obtain a degree of *buy-in* from them. Change management uses specific communication tools to reach broad audiences and address the issues and feelings of uncertainty that could sabotage any project, even when all else is progressing smoothly. It is important to have a number of strategies and protocol in place to address the needs and styles of the target audience.

Four basic strategies that are often employed in change management include: normative, coercive, adaptive, and rational. The normative strategy seeks to reeducate, the coercive strategy utilizes the balance of power, the adaptive strategy proposes that people will gradually adapt to new circumstances, and the rational strategy appeals to self-interest. However, these strategies are geared to meet the various behavioral styles of employees and do not necessarily consider the other factors of change such as the process utilized or the communication activities leveraged.

However, before you can begin to implement these strategies and protocol, you will need to identify key individuals who can give you an idea of what to expect within a given department or section. These individuals will be able to provide insight into which tools will be most useful in their area, allowing you to be able to formulate an action plan to follow. They may also be able to identify specific individuals who may require personal attention along with the specific strategy that will help ease them into the idea and process of change.

Based on the strategies and protocol in place, communication tools may include an informational intranet site, joint working sessions, random individual contacts, specific contacts with various departments or sections, and follow-up protocol; such as random contacts, online troubleshooting guides, and post-change meetings.

It is imperative to address the needs and styles of the individuals within the organization, as well as the overall business needs in the plan, because final implementation and integration is an employee responsibility. While the change itself is initially rolled out by a team of managers and subject matter experts (SMEs), it is the frontline employees who will become responsible for maintenance of the new system. By addressing their needs for security, consistency, and confidence in their ongoing role within the corporation, the overall transition will be a smoother process with low amounts of resistance.

Corporations—like individuals—have momentum. Change, when managed properly, can enhance that momentum and increase the efficiency of both parties. After all, no organization is greater than the sum of the people whose daily dedication and performance maintains the health and vitality of that organization.

> **Remember!**
>
> No organization is greater than the sum of the people whose daily dedication and performance maintains the health and vitality of that organization.

CHANGE MANAGEMENT—HITTING THE WALL

The fundamental premise behind change management is that it represents the organization's understanding and recognition of the value of individual employees, regardless of their role or seniority. It is responding to the relationship that exists between the organization and its employees.

Change management can be approached from two different directions: from the *top-down* or the *bottom-up*. The top-down approach seeks to utilize the legitimate power base of the organizational structure to reduce the resistance and gain compliance during the process of corporate change; while the bottom-up approach tries to gain a degree of buy-in from the employee base during this process.

Both of these approaches are associated with two main components of an organization—structure and culture. In order to develop a strategic plan for change management, we must first understand the corporate culture—or, in other words, how people relate to each other within the company (*the way we do business*). This insight is important as it illustrates how employees navigate the system of corporate policies and procedures to accomplish their main responsibilities and are often able to supersede their own levels of authority. It is this knowledge of corporate culture that may be used in the development and implementation of normative, adaptive, and rational strategies in a *bottom-up* approach to change management.

Next, understanding the organizational structure (or the formal chain of command) will help the change management team navigate through the barriers that are presented at the employee level. It is the chain of command that regulates the impact of change through the development of coercive and normative strategies by taking a *top-down* approach to change management.

"The power you use is the power you respect." What that means is that while the overall strategies that are developed serve as a baseline for how the company wants to deal with the resistance it will receive when attempting to implement large-scale change, we then tailor these strategies to the individual situation. Remember, change management strategies are designed to mitigate the impact of change on both the individual employee and the corporate environment.

In the wake of the announcement of impending change, people will very quickly identify themselves according to the strategy that will best suit their needs. Bear in mind that 80% of the people will use 20% of your time and, conversely, 20% of the people will use 80% of your time. The language and image—or the overall *look and feel* of both the announcement and the foundation of the corporate change itself, whether it is a deployment, new policies, or a merger—should already be consistent with the existing culture. In this situation, the adaptive and normative strategy would account for 80% of the employees. However, the remaining 20% will require the implementation of coercive and rational strategies, such as working closely with their supervisor and demonstrating that the large majority of employees have already integrated the change into their workday.

Scenario

As the change agent on a software deployment team, you have identified that many people in the corporation are happy with the way that their computers currently operate. In order to prepare for absolute resistance (people who will respond with flat refusals and hostility), we would devise a strategy for contacting the next higher up in the chain of command and utilizing the credibility of their position to sway the employee toward integration. Acceptance and buy-in will come later, so there's no reason to worry about that at this time.

Developing and implementing a rational strategy will rely on your knowledge of the corporate culture; however, these employees also tend to separate themselves into the 20% of employees who will seem resistant to change. The important difference to recognize here is that these employees are more likely to be asking for additional information before they agree to integrate the change. By providing this information, we can quickly assimilate them back into the 80% who will use only 20% of our time. The types of rational statements you can use should be based on the types of concerns they express in the initial interview.

Addressing Concerns

Argument: "We're moving to an inferior product. I won't be able to do the same kinds and quality of work anymore."

Possible response: "Actually, both products are comparable. Many people often express a personal preference for one or the other, but the company has decided to standardize its software base and after careful review of all the factors, determined that this one would become that new standard."

The importance of generating an overall strategy that uses both top-down and bottom-up approaches cannot be underestimated. It is the combination of normative, adaptive, rational, and coercive strategies that will be the most effective in addressing the needs of the entire employee base. Change management is really about managing the relationship and fostering goodwill with employees. It is about making the effort to get a degree of buy-in from the majority of employees and compliance from the remainder. We need to recognize that people have an intrinsic need for control over their own circumstances. In times of change, this control is threatened, and people will react simply as a means of regaining that control.

CHANGE MANAGEMENT—IN THE AFTERMATH

By and large, the most common platform for the implementation of large-scale corporate change is a project. The intrinsic nature of a project is to accomplish this change in as little time and with the least amount of money possible. This alone can leave team members and employees feeling as though they are being swept up in a whirlwind. In a similar way, projects have their own momentum, that is, more often than not, a world apart from the corporation itself.

With this in mind, it would be easy to understand why change management teams have their work cut out for them. Strong overall strategies and effective consistent communication throughout the process of implementation can result in minimal impact on the corporate culture and employees. Nonetheless, there will be some degree of fallout.

For instance, when the coercive strategy is utilized and compliance is the only available option, politeness and courtesy are of the utmost importance so that, in time, as the employee becomes more comfortable and versed with the new policies, they will feel that they have regained control and yet not have lost any dignity during the experience—and may even become the champions of future changes. However, in order for any future changes to be considered, we have to review those most recent changes and learn how we can improve on the process in order to ensure that new changes are implemented as seamlessly as possible.

During the process of change, we have the opportunity to learn more about the employee base from their perspective, and the technical side of the project (the tools and approaches that worked best).

From the employee perspective, the change management team needs to assess the level of empowerment and buy-in, overall comprehension of the change itself, how the employee's roles and duties were affected, and moreover, what their level of individual participation was in the process. Technically speaking, the team needs to assess the tools they used—both to communicate with the employees and to implement the change. Part of the assessment of the implementation tools used is documentation throughout the process, and the other part is to review that documentation afterward to find what worked best. For example, how often and what kinds of information was accessed from the project website?

In order to fully review the project from the employee perspective, the change management team can use tools such as follow-up meetings; online surveys and mining data from the project website; and statistics and information that were tracked during the process. The most comprehensive project review will utilize a combination of these tools in order to reach as many employees and obtain the most amount of feedback as possible.

Change, in one form or another, is an inescapable part of corporate life, as it represents the organization's ability to advance with social and economic change and its ability to respond to the needs of its employees and customers. In order to better understand the implications for future changes to the organization, we need to understand the changes we have made to date, how they affected things like employee and customer loyalty, and what the long-term return on investment will be. This understanding is simply not possible without two-way communication.

BEYOND RACI: GETTING SPONSORS, BUSINESS OWNERS, AND USER GROUPS INVOLVED

The key to any successful project, business analysis, and requirements activity is to ensure that the business is involved. As previously mentioned, setting and managing expectations is a critical part of the foundation for this involvement and keeping it going throughout the project.

However, for a variety of reasons—many of which are outside of the control and purview of the analyst or even the project team—it can be a challenge to rally the business, get the employees involved, and keep them involved. The results when this challenge becomes too much for the business analyst or the project team to overcome, are that there are increased numbers of change requests, along with schedule and budget overruns. Ultimately this means that inappropriate software features are developed, implemented, and then never used.

RACI Matrix

One of the ways in which projects work to support the business and ensure that they are actively involved throughout the project is to define the RACI matrix. In short, this matrix outlines who is *responsible* (R) for performing key tasks, who is *accountable* (A) for ensuring that it is completed, who is *consulted* (C) regarding the key tasks, and who is merely *informed* (I) of the results.

Why Some People Contribute and Why Others Don't

However, outlining roles and expectations using the RACI matrix is no guarantee of involvement by those who are impacted by the change. Again, many of the factors go far beyond the control of the change agent and even the project team. At any given time, there could be a number of inhibiting factors at work: office politics, family issues, personality issues, overloaded work schedules, feeling cheated by the company, a lack of buy-in about the need for any changes, leftover grudges from past mergers, and so on.

The point is this: the change agent has a personal responsibility to get the business engaged and involved. This is more than jotting down names in a RACI matrix, it means working with people—including all of their personal baggage—in order to ensure that they contribute. This, and only this, will ensure that the transformation effort is a success.

In addition to working with personalities, the change agent must also work to create opportunities for both the business as a whole and as individuals to contribute to the transformation. It is in creating these opportunities that the foundation is laid for this contribution to occur.

One of the main issues that arises on projects is the lack of contribution. It stems from the false assumption that simply inviting people to meetings or asking them questions in those meetings is creating the opportunity. The bigger assumption is that those people will simply show up and answer the questions being asked of them.

Many change agents then proceed to struggle with getting *the business to buy-in* throughout the rest of the transformation effort. To a degree, this again goes back to setting expectations and building a RACI matrix, but it also goes back to working with individual personalities. When dealing with transformation projects, creating opportunities to contribute takes the form of respecting people's time, asking for clear and specific inputs or feedback, and providing adequate time to respond.

To respect people's time, the change agent must ensure that each person's involvement must be limited only to those things that they actually need to be involved in. This means that when a meeting is held, the organizer must ensure that only those who will actually be contributing are invited as *required* attendees, and anyone else who needs to be informed is only on the invitation as an *information only attendee*.

Let's face it, people are not usually sitting around waiting for project meetings so that they can come and grab some free coffee and pastries (if offered). They have more important things to do in a day, especially when they also have to make time for multiple transformation projects, and that does not include attending unnecessary meetings.

Why Opportunity Alone Does Not Equal Contribution and Increase Participation

Have you ever been to a party where there are a couple of people sitting off on their own and not really interacting with anyone? They seem content simply being in the same room with people who are laughing, dancing, and having fun. However, looks can be deceiving.

Simply because they may seem content does not mean that they actually are, and it does not mean that they would prefer to sit there alone. In fact, it could be that they simply do not feel really welcome or it could be that they feel left out and/or they do not know how to participate. Of course, it could also be that they were having a great time and interacting a lot before a certain other person arrived, or even that some event occurred that caused them to shut down.

The reality is that people will either participate or will not participate for a variety of reasons that may be completely out of the control of the change agent. The opportunity to contribute to the conversation or input ideas will not necessarily get a person to contribute an idea or to share their perspective. In fact, if this kind of opportunity is presented in a meeting with heavy tension to participants who are shut down, that person may feel confronted and they could shut down even further.

Types of Participation

There are several ways in which any change agent can overcome personality conflicts, tensions, and even reach those who are shut down. First and foremost, the analyst must identify the types of participation that they expect from people before planning each meeting.

Once the change agent knows the specific outcomes from a particular meeting, it should be fairly simple to identify who and how each person can help to create the outcomes. It is important to also identify the specific decisions to be made, questions that need to be answered, or information that is needed from the session.

The change agent should be able to identify the different types of participants from among those people whom they will be meeting. There are five primary types of participants that every change agent should become acquainted with in order to be successful in their role (as shown in Figure i.1). These are: the active participant, the nonparticipant, the heckler, the hijacker, and the kamikaze.

The Active Participant

The active participant is the person who comes to the meeting prepared to contribute and get work done. They are not usually quiet, and they generally play well with everyone else in the room. This participant has no problem contributing ideas to the discussion and eliciting ideas from

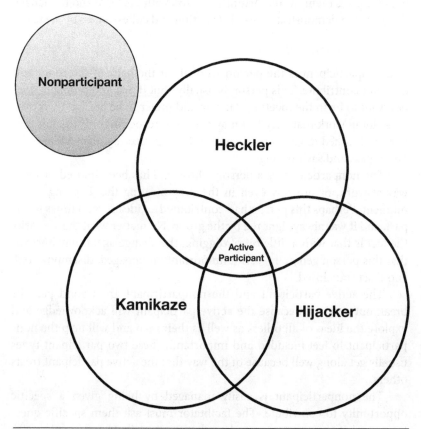

Figure i.1   Five participants theory

others. They are quite open to discussing those ideas from others in order to explore the value of the individual idea.

The active participant needs to be in an environment where everyone is encouraged and welcomed to contribute. This person is all about the ideas and merit of those ideas, and they do not take criticism of their ideas as a personal insult.

The active participant will, however, become frustrated when the goal of a meeting is not met, the work to be done in meetings does not get done, and when others are not contributing or are pushing others around. The facilitator or manager will quickly lose this person's respect if they cannot manage the other personalities in the room because the active participant wants to work on and complete the tasks.

The best way to manage an active participant is to encourage them to contribute and take the lead from time to time. This person is usually a good mentor who is well respected, so having them lead sessions when possible gives them the opportunity to work with the team and to manage the others to demonstrate the collaboration and cohesiveness of the team.

The Nonparticipant

The nonparticipant is the person who sits at the back of the room and does not contribute. This person is usually shut down and probably does not want to be in the meeting. They would prefer to be back at their own desk doing work that they feel may get them recognition. If the nonparticipant is asked to give input in a round-robin meeting, they will usually opt to pass and say nothing.

The nonparticipant is a person whose ego has been injured in some way at work, or possibly even in the very meeting that is going at the moment. Perhaps this person had contributed an idea several times in the past, and it was always ignored by the group. No matter what the scenario, the fact is that with a little bit of digging, the change agent may uncover that this person generally feels unimportant, disengaged, disempowered, and disenfranchised.

The active participant and the nonparticipant are a good pair for break out sessions because the active participant will acknowledge and explore the ideas of all others as well as their own and will help the non-participant to feel included and important. These two participant types usually get along well because of the way that the active participant treats others.

The nonparticipant is easily managed by being given a specific opportunity to contribute. The facilitator must ask them specific questions; and if they are going to be asked to contribute information at a meeting, the more notice given to them, the better prepared they will be. In this way they are more likely to contribute.

Again, round-robin discussions do not work to get this person talking, but then again, neither do general questions to the group, because they will merely let someone else answer. The change agent must address them by name and ask their opinion or a specific question if they want the nonparticipant to contribute.

Another way to get the nonparticipant involved is to approach them and get into their physical space. Then, pick up on something within their reach because it gets their attention. The nonparticipant is often a million miles away until they hear their own name being called or someone addresses them and makes eye contact. Again, the change agent must personally ask the nonparticipant to get involved in the meeting, otherwise they are just a warm body in a chair.

The Heckler

The heckler openly and loudly disputes ideas and attacks credibility. They will attack and dispute the credibility of the change agent, the manager, the solution, the organization that they work for, the department that they work in, and even others in the room.

The heckler is deflecting attention away from themselves because they do not want anyone (including the change agent) to know that they may not be able to understand or believe in what is going on. Of course this person could be bored, but an adept change agent will quickly find out if this is the case.

The heckler needs clarity, guidance, and support to help them to understand either the benefits or what exactly is going on. However, they need this to occur *outside* of the meeting. This person is good at calling others out, but does not want to be called out in return. In fact, calling the heckler out in a meeting may cause them to shut down and become a nonparticipant.

If, on the other hand, this person is just bored, the behavior will cease once the change agent has taken them aside and spoken to them about it.

The heckler needs extra attention to help them either understand or gain buy-in, so the best way to manage this person is to give them a personal demonstration, to coach and mentor them, and to ask them for questions while they are out of the group setting. Once the heckler has started feeling like he has a better understanding of the solution, or at least has the feeling that the change agent is a person whom he can trust, the heckler will begin to contribute in meetings.

The Hijacker

The hijacker is the person who tries to take over control of the meeting. Typically, this person feels as though they should be at the front of the room instead of the facilitator, so they will make every effort to take control of the room by directing the conversation, steering the agenda off course, and having side conversations.

It is simple: the hijacker does not respect the authority of the manager or the change agent because they want to be in their position. This person usually feels a degree of supremacy over the others in the room or on the team, but they could also be feeling jilted. Perhaps in some way, the hijacker feels passed over.

The hijacker needs an ego boost, public recognition, and public attention. Ironically, what this person needs most of all is to feel as though the change agent is their ally. Remember that the hijacker may be feeling passed over, and they want recognition, so if the change agent gives it to them, they will respond positively. This recognition does not always work on the first few tries, but keep going, and it should have a positive effect once trust has been established.

The best way to manage the hijacker is to give them some time to speak during the meeting, give them public praise, offer them time after the agenda items are covered to discuss their burning issues, ask them to facilitate when there are break-outs or the facilitator is going to be away, and to set and control the limits on off-topic discussions.

The Kamikaze

The kamikaze is the most aggressive of the personality types. They assert their opinions and choices over others and tend to try to dominate everyone.

The kamikaze will go to the death (firing, loss of relationship, etc.) to be right, even when this decision is not logical or founded in facts. The kamikaze often rejects the consequences of their actions. With this personality type, the adage *we judge others by their actions and ourselves by our intentions* is absolutely true.

They believe they are immune to consequences when they believe that their intentions are just. However, that doesn't stop them from passing judgment on others and enforcing consequences as they deem fit. Conversely, when others attempt to impose any consequences on them for their behavior, they are likely to view those consequences as a personal attack. They will blame others around them for the conflict because they are simply exercising their right to choose and because they have judged their intentions as just.

The kamikaze is a difficult personality to manage, and they require the close guidance and leadership of a person who is older and wiser; someone that they feel a great deal of respect for. This person is often the only person who can talk to the kamikaze and coach them, let alone tell them what to do.

CREATING THE RIGHT CONDITIONS INCREASES PARTICIPATION

The first few meetings may not be very productive, despite best planning and efforts, until the four main participant types are identified. These participants may make it difficult to manage the room and to accomplish any significant amount of productive work. This being said, there are ways in which the business analyst, acting in the role of a facilitator, can set up and structure a meeting so that people are encouraged to attend and contribute.

There are a few key things that any change agent can do in order to increase the likelihood of success in getting people engaged, involved, and actively participating. These include: conducting routine informational activities, creating input funnels, and being a facilitator.

Informational Activities

There is not a human being alive who does not need to feel important in some way. This need is so strongly ingrained that it means a person's job, and how well they do this job, becomes a part of their personal identity.

Further, when something new comes along and this something new is perceived to threaten his or her identity, the person must have an opportunity to provide input into the new *situation*, to have their concerns heard and questions answered. Transformation efforts that do not take these factors into consideration are doomed, even if those efforts were to deploy a solution made of gold, because the business will revolt against both the project and the solution.

Successful change projects must start with a high-level set of activities that provide information to the executives and then works to get those executives involved. Once this involvement has started, the project team will begin to disperse information to the business and customer communities in order to make them aware that changes are coming.

While informational activities at this point are not a heavily intensive process, these activities do enable people to prepare themselves mentally and emotionally for impending change. These informational activities must be able to provide basic information about what the business and its customers can expect, who will be impacted, and how. In addition, this early information should also provide details about the expected participation from each impacted group, where they can find more information, and how to contact the project team with any questions, comments, or feedback.

Input Funnels

A well-planned transformation project contains both outgoing and incoming communication channels. Those channels include combinations of informational and input activities (outgoing channel) and feedback funnels (incoming channel).

While the outgoing channels disperse information from the project team, the incoming channels provide opportunities for the business and its customers to respond and provide their thoughts and inputs into the new solution. In this way, the business is provided with very specific and direct methods and opportunities to contribute throughout the transformation.

These channels and funnels help the business and its customers to feel both heard and important, and ultimately increases their participation levels. However, it is not enough to hear what is being said, it is also critical that those inputs are then incorporated into the solution where and when appropriate.

Many project teams make two common mistakes when they plan change management strategies. Often they will only consider the cost of the communication channels or they only consider outgoing channels.

When the project team only considers the cost of communication channels, this mistake is made purely as a cost reduction measure. When cost is the primary factor in selecting and implementing communication channels, the key message that is communicated to the business community is that their input is not valued. Unfortunately, these project teams limit the ability for two-way communication and reduce the ability for the business community to participate. In addition, the channels themselves are very one-sided because they are intended to disperse information out from the project.

When the project team mistakenly considers only outgoing channels, this mistake is made under the erroneous assumption that requirements elicitation is when stakeholders and users will have an opportunity to provide input. Unfortunately, this is a fallacy because many requirements are often generated by working exclusively with SMEs. All too often, those SMEs make decisions based on their own personal experiences or opinions rather than considering those from the team at large.

By waiting for the requirements activities to begin, it is far too late in the project to begin collaboration. This timing factor makes the buy-in process much harder and actually increases the likelihood of changes to scope as well as a lower adoption rate.

The Change Agent as Facilitator

Every change agent must know and understand the role of a facilitator before any of the real activity can begin, because it will be one of their primary responsibilities throughout the project. The role of the agent is to ensure that those impacted by change will actually adopt it. This means that in any given meeting or session that they will conduct, the change agent is more than an attendee, and he or she is more than a participant.

It is important to remember that *buy-in* means *believe-in*. There is a lot of energy and excitement in starting a new project. The best way to build buy-in from people is to get them excited, get them involved, and show them how to carry it forward.

When people are informed that change is coming, it is new and exciting, yet scary all at the same time. The change agent must anticipate and overcome people's fears of being replaced or phased out or suddenly being seen as incompetent by collaborating with them.

Many times, the change agent can identify a single person or a group of people who are being obstinate and have blockaded the way for change and the new solution, because they do not believe in it. By working with this person or group, understanding their needs, and finding ways to meet those needs without changing the project course, the project will find a new champion.

The loudest and most outspoken adversary can become the project's biggest champion. When this occurs, this person supports the initiative and it becomes a grassroots movement. The project is now beyond buy-in when this happens because whether or not others actually like the champion does not matter.

What does matter is that people know how the champion behaves. When the champion is happy with something, they will let everyone know, and the life of those around them is infinitely better. On the other hand, when the champion is angry, everyone is going to know about it and they will prevent others from participating because they want people to support their position, and they become a divisive force.

Knowing and understanding the types of participants is crucial to being successful in hitting this tipping point. It is equally important for the change agent to know and to understand that their role is more than merely to distribute surveys and to analyze data. The full role of the change agent is to sponsor, support, and promote the changes being made. This fundamental change in attitude will ensure that everyone is encouraged to collaborate in an innovative space and environment where everyone's contributions are valued and important to the success of the project.

Understanding the role of the change agent as outlined above is really only the start of the positive project experience. In order to be successful with encouraging buy-in, the agent must be a good ambassador, psychologist, psychiatrist, mediator, therapist, customer service representative, host, salesman, marketing expert, and negotiator. It is the elements and combinations of each of these roles that help to increase buy-in and to move both the business and technology teams past the RACI matrix into the role of an actively engaged participant.

REFERENCE

1. *Monty Python and the Holy Grail*; 1975.

The Barriers to Change

One of the first things that people need to know about change in order to be successful in implementing organizational change management strategies, methods, and approaches is that the very nature of change creates conflict. As depicted in Figure 1.1, that conflict is both internal and external.

Someone once said, "Change is both exciting and scary." It is this paradox that creates the conflict associated with change. Change impacts how people feel about themselves and the world around them. Change is personal, even in an external environment like work.

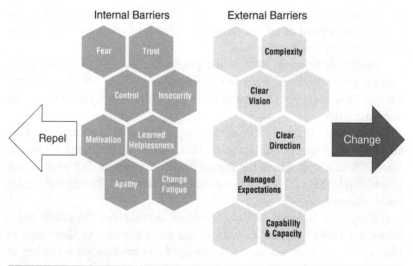

Figure 1.1 Barriers to change

The personal nature of change means that the idea of impending change evokes reactions that expose fear, trust and control issues, apathy, insecurity, motivation, learned helplessness, and the level of change fatigue. However, the openness and willingness to embrace change also depends on a variety of external factors such as complexity; clarity of the vision and direction; management of expectations; as well as the capability and capacity for the specific changes.

In combination, these factors create a double wall that becomes a very solid barrier to implementing change. In fact, if these barriers are not addressed, it almost doesn't matter what changes are being implemented, the adoption rate will be lower over a longer period of time than is necessary and, perhaps even, sustainable.

The resistance to change increases when both the internal and external barriers (fear, trepidation, nervousness) are not addressed. In combination, these barriers decrease the likelihood that change can occur successfully. Therefore, any organizational change initiative must address these factors in its strategy and plan by leveraging very specific tools and techniques.

Let's talk about the 80/20 Rule (or the Pareto Principle) in the context of organizational change management because it's an important thing to keep in mind as changes are planned, implemented, and managed—especially when talking about how people react to change and the internal and external barriers that will exist throughout the process of change.

Ordinarily, the Pareto Principle states that roughly 80% of the effects come from 20% of the causes.[1] Within the context of organizational change management however, let's say for the sake of argument, this rule can also be applied as an indicator of the types of resistance that will be evident on any one project.

Look at it this way, 80% of the people are actually going to accept change. They will accept it, in part, because people have a tendency to *roll with the flow* and, in part, because others are doing it (very few people like to be the loner).

That means that only 20% of the people impacted by change are actually resistant. This does in fact mean that 80% of the resistance is coming from 20% of the people as the Pareto Principle suggests. But what is also interesting is that of those 20%, a further 80% are actually only asking for more information.

What this suggests is that communication plays the single most important role in organizational change management. As illustrated in Figure 1.2, well-planned and well-managed communication is the key to engagement, collaboration, and adoption of change.

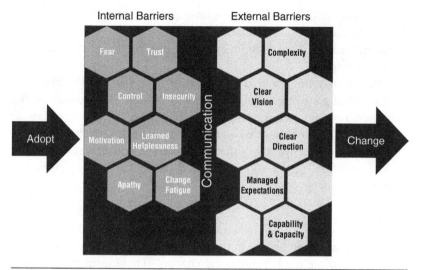

Figure 1.2   Bridge to change

INTERNAL BARRIERS

Fear

Fear is exactly what it sounds like. It is the fear of the unknown that arises at the mere mention of change. It is the natural gut response to any type of change announcement and is obvious when people start asking questions like:

- What are you asking me to do?
- How will this impact me?
- How will my current responsibilities change?
- Are we sure this will work?
- Who decided to make this change?

How Does Fear Impact the Ability to Change?

This doesn't mean that fear is the end of the change. In fact, it can be quite the opposite when addressed correctly. Since fear is a natural response, it is important to understand how it can impact change if and when it is not addressed.

Again, the fear of change is really the fear of the unknown. The best way to combat the fear of the unknown is to give people time to overcome the initial shock—show people what life will be like with the new changes in place and give them time to get comfortable with them.

Specific techniques that can be leveraged to combat fear include up-front communication, visualization, quick wins, mentoring, and training sessions. Each of these techniques will be discussed in further detail in Chapters 4 and 10.

People may still adopt changes if they have fear, but the likelihood of those changes being sustained is greatly reduced. The reason for this is simple. Fear blocks our cognitive ability to hear, see, learn, and retain valuable information. In other words, it can make the activities you have for training ineffective and pointless because it will actually lower the rate of retention for the information and flatten the learning curve (see Figure 1.3).

Trust

Trust is the ability to believe in the new products and processes, but also in those who are implementing the changes. This means that in order for people to step out onto that ledge and adopt changes, they must first trust that getting onto the ledge is the right thing to do under the circumstances, trust that the ledge will hold them, and trust that they are being given sound advice.

How Does Trust Impact the Ability to Change?

Trust impacts our ability to change for just those reasons cited in the previous paragraph. Based on the reasonable belief in any combination of those factors, people will generally at least try to step out onto the ledge. However, if they don't trust the leader or other person who may be giving them the advice, it's not going to happen. After all, they have to be out on that ledge with that leader. Why risk being pushed off?

Specific techniques that can be leveraged to build trust include up-front communication; clear direction; creation of a strong vision

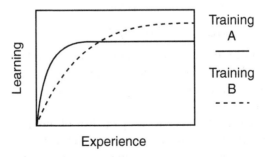

Figure 1.3   Learning curve comparison

statement; road mapping; Kaizen events; determining a solid communication architecture; hosting forums, hackathons, and skunk works; quick
wins; mentoring; and training sessions. Each of these techniques will be
discussed in further detail in Chapter 9.

The fact is that people are more likely to leave than adopt changes
when there is little to no trust. Those who do adopt changes often aren't
doing so because they believe in them; rather, they are apathetic.

Control

Simply and obviously put, this is the level of control that an individual has
over what they do on a daily basis. In *self-determination theory* (SDT), this
is referred to as the fundamental need for autonomy. It is the reason that
so many immigrants came to the Americas to start new lives for themselves and their families.

How Does Control Impact the Ability to Change?

In terms of change, people are more willing to adopt proposed changes
if they feel as though they have some sense of control over what is being
changed, as well as how and when it is changing. However, in the long
run, the level of personal control does not mean that people will not adopt
the changes. It may mean that they adopt some of the changes and implement work-arounds to skirt others.

Specific techniques that can be leveraged to empower people to collaborate through changes include hosting forums, hackathons, and skunk
works; conducting quick wins; and leveraging communication architecture to accept and incorporate feedback.

As with trust, the level of control felt by people in implementing
changes may have little impact in the ultimate adoption of the changes,
however, again, they may be more likely to leave or introduce their own
work-arounds. People will take control where they feel that they can get it.

These work-arounds are really their way of saying, "I'm not doing
what you told me to." Unfortunately, one of the other problems with a lack
of a sense of personal control is that people will actively work to sabotage
the project. They will gossip, skip important meetings, not contribute,
speak out against the project, and actively encourage others in a corporate
anti-project mutiny.

What that really means is that these guys are the 20% of the 20% in
the Pareto Principle. They need more, they need it now, and they need
recognition for it. Otherwise, good luck with that change thing!

Apathy

Apathy is a lack of concern and complete indifference to organizational changes being introduced. It is significant in that it is a direct indication of the level of buy-in and trust that people have for the company and its leadership. This is often accompanied by the laissez-faire *I-get-paid-to-be-here* attitude.

How Does Apathy Impact the Ability to Change?

Change apathy means that people will neither work to support change, nor will they work against it. On its own, apathy is a sense of indifference in that people just don't care enough to try, and that signifies that they are less likely to embrace change or collaborate on problem solving when issues arise during the change process. They are more likely to quit or go back to the old way of doing things or simply ignore the directives to change. According to SDT, the fundamental underlying need that these people have is relatedness—in that they need to feel as though they are connected to and are an important part of the team and changes around them.

Specific techniques that can be leveraged to decrease apathy are really the same as they are for control because people become apathetic if they feel as though they don't count and are not important in the process. Again, these include hosting forums, hackathons, and skunk works; conducting quick wins; building networks and practice communities; and leveraging communication architecture to accept and incorporate feedback into the changes. However, it also includes celebrating and recognizing peoples' contributions to the success of change.

Apathy doesn't mean that people won't adopt the changes, but it does mean that those changes won't be sustained. This is exactly why different projects are initiated years later to reintroduce similar changes to the company, and why a review of the ongoing processes and procedures will look very different to what was implemented by the project team.

Insecurity

Insecurity is the feeling of being anxious, uncertain, or timid. In terms of change, insecurity is closely related to fear in that it is the fear of the unknown that arises at the mere mention of change and is the natural gut response to any type of change announcement. However, it only arises when the person or people impacted by the changes have self-doubt about their own individual or collective capabilities and capacities to adopt the changes.

Insecurity is obvious when people say things like:

- That will never work here.
- Those companies are not like us.
- I could never do/try that.

How Does Insecurity Impact the Ability to Change?

Insecurity impacts change given that it limits peoples' ability to personally connect with the changes being made and often leaves them feeling left out. As a result, they become disengaged from the team and do not contribute or collaborate. According to SDT, the fundamental underlying need that these people have is to feel competent.

Therefore, specific techniques that can be leveraged to build up the confidence of those impacted by change include mentoring, conducting quick wins and workshops, and celebrating successes. Remember the analogy about the ledge? Well, these people need to know that they can stand on the ledge without falling. What they really need is to learn to trust in themselves, regardless of the situation they find themselves in.

People who are insecure will adopt proposed changes, but they will do it at their own pace unless their confidence in their capability and capacity is improved. As with fear, their ability to learn is impacted by insecurity. Feeling insecure often makes people give up or prevents them from hearing, seeing, and learning because their perception is colored by a negative view of themselves.

Motivation

Motivation is the internal drive or impulse to perform a given task or set of tasks. SDT describes motivation as the rationale that drives the choices that people make of their own volition, without any interference or influence from external sources.

How Does Motivation Impact the Ability to Change?

While motivation may not always be a direct result of external factors and sources, those factors can impact a person's motivation (as we will discuss in the section *Learned Helplessness*). They can also impact a person's fear, insecurity, apathy, and levels of control. Thus, one could say that motivation to change is impacted by these external factors about as much as it is by internal factors.

Specific techniques that can be leveraged to increase a person's level of motivation include up-front communication; visualization; conducting quick wins, workshops, mentoring, and training sessions; clear direction; creation of a strong vision statement; road mapping; Kaizen events; determining a solid communication architecture; hosting forums, hackathons, and skunk works; celebrating successes; and leveraging communication architecture to accept and incorporate feedback.

Because motivation is closely tied to fear, control, insecurity, apathy, and trust, it has a direct impact on peoples' willingness to adopt changes and the lengths of time it will take to implement them. Consequently, it will also have a direct impact on the ability of the company to sustain those changes once they have been implemented or deployed.

Learned Helplessness

Learned helplessness is similar to apathy in that it is an attitude of indifference. However, it is a learned attitude that is the direct result of being dejected time and again.

Learned Helplessness

When elephants living in captivity are small and young, they are tied to bushes by heavy ropes. After struggling in vain to escape the ropes, the elephants give up. When they have grown up, they hold onto the belief that they cannot possibly escape the bonds of the rope and the bush that is no longer a match for their sheer size and strength.

However, since the elephant has long since given up, she will not try to pull the bush out of the ground or break the ropes.

How Does Learned Helplessness Impact the Ability to Change?

Learned helplessness can impact people's ability to change because when they do not feel as though they can change or that change is just too daunting, they are less likely to participate in the initiative. What does this look like?

This means they do not fill in the survey about what is important to them, they do not attend the town hall meetings, they do not provide input into requirements, and they do not respond to e-mail requests. After all, why bother? In their minds, it cannot be done, so they are not going to show up and help the project fail.

There are numerous incidents of learned helplessness in companies where the employees and customers are disempowered. When it comes time to change, as with apathy, people probably won't care enough to try anything new and will be less likely to embrace change or to collaborate with others. Again, they will be more likely to quit, revert, or ignore the change directives. According to SDT, the fundamental need is related-ness—meaning that they need to feel as though they are an important part of the change.

Specific techniques that can be leveraged to decrease learned help-lessness are the same as they are for both apathy and control. Again, these techniques include hosting forums, hackathons, and skunk works; conducting quick wins; building networks and practice communities; and leveraging communication architecture to accept and incorporate feedback into the changes. However, above all else, it includes actu-ally incorporating the feedback, celebrating successes, and recognizing peoples' contributions. Learned helplessness will not prevent people from adopting the changes, but those changes will likely not be sustained.

Change Fatigue

Change fatigue is the mental and emotional exhaustion that comes from too many changes being implemented over an extended period of time. In general, people need time to rest and develop new norms under the changes before moving on. Think of it in terms of a cross-country road trip or a flight from Los Angeles to Bangalore—at some point, you're going to need to rest and adjust so that your body and mind can catch up to a new reality.

People experiencing change fatigue often remark that there is *always some new project going on* or that they want things to slow down and *stay the same for a while*. This sentiment may be held by several groups within the organization, but not necessarily the whole organization. When there is a vast difference between sentiments expressed by groups across the organization, it can be an indicator that more change management efforts are needed because, ultimately, it illustrates that some people are not con-sistently seeing the vision and the path forward.

How Does Change Fatigue Impact the Ability to Change?

Change fatigue impacts the ability to change since it can actually make people apathetic, feel a loss of control, feel insecure, and lose trust in management. None of these are good places to be. However, this can be overcome by addressing not only these individual issues, but also by stag-ing change in phases; focusing communications on the long-term, big vision; and celebrating successes.

Is there a way to motivate people who are already beleaguered and suf-fering from change fatigue? Ultimately, the specific techniques that can be leveraged to combat change fatigue are the same as those to combat each of the other internal factors. But there is one big thing that can be done to decrease change fatigue, other than implementing in phases—that is to turn change efforts into a grassroots movement. In doing so, the stakeholders themselves increase their own motivation to change and work to find ways to maintain momentum throughout the effort.

EXTERNAL BARRIERS

The next thing that people need to understand about change in order to be successful is that it is influenced and impacted by a variety of external barriers. These barriers can directly affect (either positively or negatively) the internal barriers, which may prevent people from readily adopting change and the long-term sustainment of those changes. These barriers include:

- Complexity of the change
- Lack of clear vision
- Lack of clear direction
- Mismanaged expectations
- Lack of organizational capability and capacity

Change Complexity

Change complexity refers to the number of factors and dimensions that are going to be changed, as well as the types of those dimensions. For example, a single factor would be the number of stakeholders impacted by the changes; and a dimension would be organizational redesign.

How Does Change Complexity Impact the Ability to Change?

It almost goes without saying that the complexity of the changes to be made impacts the ability to change—almost. Yet, in spite of this, far too many projects and operational changes are implemented with almost no discernible organizational change management strategy, let alone one that takes complexity into account.

Make no mistake, this is a *big* problem. Organizational change is like any other change, complexity *matters*.

Consider This

You take your car in for an oil change before heading out on that great road trip to see the Grand Canyon. You expect it to take an hour and have planned your day around that. Unfortunately, the mechanic puts it up on the hoist and promptly reports to you that it needs a major engine overhaul. Not in the budget. Not in the plan. And totally kills your day.

Specific techniques that can be leveraged to assess complexity include business criticality, audit and functional complexity, business and stakeholder readiness, organizational preparedness, as well as the numbers and types of stakeholders, functions, processes, and business units impacted. These will be discussed in greater detail in Chapter 10.

Lack of Clear Vision

A clear vision statement defines the outcomes of the project or the changes in terms that give the stakeholders (employees, customers, vendors, and partners) a goal to be achieved. Vision begs the question: *where will we be in one or two years after this change has been implemented?* It is not designed to sell the change to the stakeholders, but rather to enlist their help in building it.

How Does the Lack of Clear Vision Impact the Ability to Change?

Every change, no matter how big, small, or complex, requires a vision for what the organization will look like once the changes have been successfully adopted. Without a clear vision or goal, change becomes a chaotic churn for all of those involved. Worse, it is left open to suggestion and interpretation by those trying to implement the change—any chance of buy-in dwindles or goes right out the window.

Lack of Clear Direction

The lack of clear direction is closely related to the lack of a clear vision. Direction is set and maintained by the leadership and management team of the organization. It is this direction that keeps every one of the stakeholders moving at the same pace—toward the same goal.

How Does the Lack of Clear Direction Impact the Ability to Change?

The lack of clear direction stunts the organization's ability to change because stakeholders are in chaos and moving at different rates, in different directions—some aren't even moving at all. It is imperative that leadership establishes and maintains a clear vision and direction, regularly communicates them outward toward the stakeholders, and also holds routine check-ins to obtain the perspectives of the impacted stakeholders.

Mismanaged Expectations

Expectations are the beliefs that are held by each stakeholder and leader about how the changes will progress, be deployed, and be adopted. Far too often, these are implicit assumptions rather than explicit agreements. However, whether implicit or explicit, they require management from setting to changing expectations.

How Do Expectations Impact the Ability to Change?

The ability to change is impacted when expectations are implicit and unmanaged by the leadership and the project teams. Although many stakeholders will come in with their own preconceived assumptions and expectations, it is the job of both leadership and the project team to discuss and set explicit expectations. Ultimately, all of the stakeholder expectations must align with what is going to be achieved by the change initiative.

The single most important technique that can be leveraged to set and manage expectations is the development of communication architecture. It is this infrastructure and planning framework that will guide the management of expectations throughout the process of change from cradle to grave.

Lack of Capability and Capacity

Capability and capacity are often spoken of in the same breath. This is because they are interrelated. Capability is the skill and know-how that is required by an individual or an organization to accomplish a specific task or activity in the achievement of a goal. Capacity, on the other hand, is the breadth or amount of that same skill and know-how.

In terms of change, capability and capacity means that people and organizations need to know *how* to change, and they also must have enough of the skills to work through with change. For example, within an organization that is implementing new software, capability to change could be having the knowledge of the new software applications, and capacity could be the number of trainers hired to ensure that everyone gets the required training to become comfortable using it.

How Do Capability and Capacity Impact the Ability to Change?

Capability and capacity play a pivotal role in change because they have a direct impact on an individual's ability to adopt the new policies, procedures, tools, and organizational structures that are implemented with the changes.

Specific techniques that can be leveraged to increase individual and organizational acceptance of change include direct and cross-training, quick-win projects, workshops, practice communities, and to a certain extent, celebrating successes. It should be noted here that there is an emotional and cognitive link to hearing positive messages that reaffirm a person's capability with their individual ability to improve on that specific skill.

Effectively, organizations are a collective of individuals. This means that change happens at a personal and individual level, and that change then permeates the organization as a result.

REFERENCE

1. Bunkley, Nick (March 3, 2008). *"Joseph Juran, 103, Pioneer in Quality Control, Dies,"* New York Times.

2

The Reasons for Change

Ultimately, there are many reasons that organizations undergo change. These reasons stem from both good leadership and bad management, both sound and flawed business decisions, changes in leadership, changes in vision, changes in organizational size, competition, and customer demand. In fact, there are as many root causes that drive the impetus for change as there are organizations across the globe.

That being said, these numerous root causes can be boiled down into a few all-encompassing reasons for organizational change. The most important reasons include what I call *burning platforms*, which are initiatives for continuous improvement, to facilitate growth, and during mergers and acquisitions.

BURNING PLATFORM

The burning platform is a euphemism for a state of emergency within the organization. This emergency can be anything from sharp drops in revenue, market share, or profitability; the threat of legal action; and/or poor business planning.

These types of emergencies are what create the situations in which organizations must change in order to remain viable. Effectively, a burning platform is a situation that threatens the health and life of the organization.

How Does a Burning Platform Impact the Ability to Change?

Unfortunately, a burning platform can create a state of panic among executives. This panic can lead to one bad management decision after another, and bring about chaos in the organization. Change management is often lackluster at best, but more often than not, it is completely nonexistent.

To make matters worse, the people within the organization often reject the decisions and ignore directives made during these times of panic. They do this for a very specific reason—the loss of trust in the executive. It is not that they get a front-row seat to watch the panic that causes the loss of trust—just the opposite, because we all panic in chaos so that's an anticipated reaction.

The biggest factor in the loss of trust is the sharp decline in communication and engagement. As executives sinks into panic, they increasingly shrink the communication circle and make decisions without the people who are directly impacted by them.

Once trust is gone, it is not only extremely difficult to regain, it actually creates a sense of apathy among the people within the organization. In other words, without trust in leadership through communication and engagement, people within the organization have little to no interest in implementing the changes required to salvage it.

A burning platform can effectively lead to a downward spiral of the organization that decimates it. That being said, it doesn't have to end this way. In fact, leveraging the power of engagement and communication with specific organizational techniques can turn the entire organization into a powerhouse of fierce change champions.

The specific techniques to leverage in the event of a burning platform can include Kaizen events, hackathons and skunk works, as well as visioning and road mapping. These are techniques for engaging people in change and empowering them to become the champions of change. However, these techniques cannot, and should not, be applied without the corresponding tools of communication architecture, specific communications, and above all, leadership.

Suffice to say that a burning platform doesn't mean the end of an organization—it is simply a critical emergency situation (much like the difference between the Titanic hitting the iceberg and say thirty-three Chilean miners getting trapped underground). When leadership pulls together and increases engagement throughout the situation, the organization cannot only be salvaged, but it could actually come back bigger and stronger.

> ### The Lego Comeback
>
> Many of us grew up playing with Lego's trademark brightly colored building blocks. However, by 2004, the company was in dire straits and struggling with both consumer demand and profitability; enter Jorgen Vig Knudstorp in the role of CEO.[1]
>
> Knudstorp not only mandated fiscal responsibility and accountability, he also engaged the brand's fan base in the creative direction of the company's products. In doing so, he engaged the very people who would be impacted by the loss of the product.
>
> To date, the Danish toy maker has an estimated 26 movies (including full-length features and shorts) and approximately 38 video games. By the end of 2015, Lego reported revenue of 27.9 billion DKK (Danish Krone) or an estimated 42.6 billion in US dollars. It's fair to say that is a pretty impressive and substantial comeback. (Based on conversion rates in April 2016.)[2]

What we learn from the Lego story is that leadership and engagement are critical game changers when it comes to facing a burning platform. Both of these elements play a critical role in how people take up the cause of changing the organization. But the truth is that engagement is not possible without communication.

Lego leveraged social media and events such as the Adult Fans of Lego Convention (AFOLCon) in the United Kingdom and a designer recruitment workshop to spread their message and engage their consumer base.

If you think about it, start-ups that leverage crowd-sourcing are doing something similar. Think of the limited budget, knowledge, and experience of the entrepreneur as being the catalyst that creates the burning platform. The limited budget creates a time-crunch (because time is money), the limited experience creates quality and product viability issues, and their inexperience can create a situation where the leader is so busy putting out fires (and not necessarily the right fires in the right order), their business can literally crash and burn before it's even fully off the ground.

By engaging their consumer base through communication and then inverting the feedback they received into product changes, Lego addressed the underlying needs that people have for relatedness and autonomy. In a situation with a burning platform, it can be difficult to trust and take chances implementing consumer suggestions, but this can prove successful when managed correctly and the right suggestions are implemented.

Therein lies the secret: people usually want (and ask for) a sense of *some* control (not necessarily total control). They want to feel as though they are important and matter, and that comes out of the need for relatedness and competence.

Managing the changes associated with a burning platform is crucial. Leveraging organizational change tools and techniques on that platform is the critical differentiator between success and failure.

CONTINUOUS IMPROVEMENT (EVOLUTION)

Continuous improvement is the never-ending and ongoing mission of organizations to improve some aspect of itself. That aspect could be anything from service or product quality, market share, profitability, time to market, social responsibility—and the list goes on and on.

How Does Continuous Improvement Impact the Ability to Change?

In order to improve these aspects, the organization creates a strategy and then undertakes project initiatives to execute the strategy. This effectively means that they *desire* change and actively *pursue* it. Effectively, continuous improvement is the epitome of change.

However, that does not mean that change is any more embraced in this instance than it is in any other. In fact, one of the drawbacks of continuous improvement can be *change fatigue*. As we discussed in Chapter 1, change fatigue is the mental and emotional exhaustion that people in the organization acquire when they feel as though there is no end to the changes being made.

That being said, continuous improvement is not synonymous with change fatigue. In fact, by leveraging specific tools and techniques, continuous improvement can be embraced and celebrated by the organization.

Let's face it; we have been designed to create an ever-advancing society. If we hadn't been, we would still be living wild—maybe using stone tools or living under the Roman Empire. The reason we aren't is because we didn't get to some point in time and all agree that we had enough of the right tools, were happy enough, and felt that life was just good enough as it was. We constantly worked to improve the tools we used to better our lives and those of others around us.

Change is literally a part of our DNA. And so is the need to relate to others, the need to be competent, and the need for autonomy. These underlying needs are what we can satisfy by leveraging very specific tools and techniques that enable and empower people to maintain momentum, in spite of multiple continuous improvement efforts that might otherwise lead directly to change fatigue.

Facebook

The best example of continuous improvement comes from one of (if not the most) popular online social media sites: Facebook. On the surface, Facebook still looks similar to what it did in the early days, but there are stark contrasts to its business model, as well as to its platform. The change has been both organic and revolutionary.

Facebook got its start as a platform to connect college students on campus. By 2004, Facebook was growing in popularity outside those campuses and by 2009, Compete.com declared Facebook to be the most popular online social network. It literally went from a couple of guys hacking code in their college dorm to a publicly traded phenomenon.

The interface and functionality of Facebook is dramatically different than it was ten years ago.[3]

In spite of periodic outcries to changes and threats to *close my account*, it remains the go-to platform for people to connect with others across the globe and to connect with businesses. One of the things that Facebook has done to help people accept the constant stream of change is primarily two-fold. The first is that Facebook allows people to customize their own pages (their *wall*) with things that are important to them and to share them outward. The second is to incorporate suggestions from the people that utilize the platform into how it functions.

Case in point: The *Like* feature has become a more diverse emoticon selection so that people can share how they actually feel about an article or post that appears in their timeline. We literally asked for it. The other thing we asked for was the *I'm okay* feature that allows people in disaster situations to declare themselves as *safe* so that their friends can be reassured that they were not impacted by the event.

What we can learn from Facebook is that people want some sense of control over the changes being made. This speaks again to the underlying need that people have for autonomy—allowing people not just to post and share the things that are important to them, but also allowing them to customize their personal *wall*. By incorporating suggestions made by the users, Facebook is creating an ongoing platform for engaging people in the broader audience as to the changes being made.

People help to support what they helped to build. That means there is an emotional connection to the things that we build.

Sometimes, that emotional connection can make people resistant to change because they want to hold onto the success they have had. That speaks to the need for feeling competent. However, this can be overcome by showing people how to be successful in the new model. Let's face it, people like success and are afraid of failure. So by having a mini-demo pop-up when changes have been made, Facebook is overcoming that fear quickly and connecting people to the new changes.

PayPal

What began as a company selling encoding and decoding (also known as cryptography) services has become one of the world's most preeminent online payment solutions. After several years of continuous improvement (trial and error), PayPal managed to find its calling as the go-to online payment system that has become ubiquitous with money transfers and payment solutions.

It is noteworthy that somewhere in the midst of all of these changes, they were forced to overcome a massive user fraud scheme that almost destroyed it before it really began. (I actually remember getting this scam e-mail.) Ironically, in some way, it seemed to help to propel them to where they are today. At the time of getting that e-mail, many people had never heard of them.

The ongoing transformation was a full-time effort and the company regularly weighed the pros and cons of remaining the same in the face of adversity or changing business models. They changed—and changed more often than many companies would dare to dream of. "But ultimately, their flexibility proved to be a major asset. Despite being founded in 1998, PayPal was swift enough to change course in time to go public in 2002 and later get bought out by eBay for $1.5 billion."[4]

What we can learn from PayPal is that continuous improvement doesn't have to be a burden on an organization's employees and customers. In fact, by changing to adapt to the ever-changing barrage of consumer needs, a company can stay relevant. Effectively, this hits upon the same need for autonomy that Facebook satisfies.

When an organization changes through continuous improvement, it is really adapting to that shift in consumer need and mind-set. By keeping up with the changes, they are addressing autonomy head-on. That's because autonomy is the ability to choose for one's own self. While that may sound like it's just about choosing colors, it's also about dictating what is needed to the company and then having the company implement that feedback into the product or service that it offers.

But the truth is that it is more than that. By incorporating consumer feedback into the products and services through continuous improvement efforts, the company is also aligning itself with the consumers that it serves and this in turn, addresses the consumer's need for being related and connected. It demonstrates, "I heard you," and that is the most powerful thing that an organization can say in the face of the constant change that comes from continuous improvement.

Managing the change component of continuous improvement is critical in order to control change fatigue. The best way to manage it is to create a strong vision and leverage communication to help people to keep their eye on the prize.

FACILITATE GROWTH

Many organizations transform their business operations to facilitate growth. This can be growing in overall size, growing revenue, or market share, or increasing the scope of products and services.

In preparation for and during the process of growth, many things will change. These things can include technology, employee head count, processes, and branding. Consequently, managing change when considering the growth of an organization is very important.

As with continuous improvement, changing the size of an organization usually involves various projects that are designed to implement the incremental increases. Transforming the size of an organization impacts the ability of its employees to change because it can increase the employees' internal barriers (their individual personal feelings) of insecurity, fear, incompetence, and loss of control. As discussed in Chapter 1, these feelings, in combination with the external barriers of vision, direction, complexity, expectations, capability, and capacity, are closely tied to an individual's motivation to change.

Example of a Business that Changed as a Result of Growth

An energy company had recently revised its long-term strategic plan and created a goal for growth. They soon realized that there were a significant number of projects that needed to occur in order to achieve the goals established in that plan. Management got together and determined that they had almost no track record for executing so many large-scale projects all at once.

Obviously, they wanted each of the projects to be successful. They realized that in order to ensure success, they needed to change their relationships with vendors and the way in which they executed projects.

They established a quality center for testing and changed the way that they conducted technology requirements activities.

In the previous story, the company changed its operating structure to prepare for and accommodate the growth that they wanted. By the time they were ready to initiate their first enterprise-wide project, they were ready for success.

SCALING BACK OPERATIONS

The opposite of facilitating growth is the scaling back of operations. Many organizations do this when there is a need to reduce overall size, locations, employees, or the scope of products and services offered. Typically, this is due to financial constraints, but it could also be to rebrand, reposition, or reorganize operations.

How Does Scaling Back Operations Impact the Ability to Change?

Scaling back operations makes everyone in the organization nervous and afraid. In the back of everyone's mind is the big question: "Am I still going to have a job when this is all over?"

Obviously, when people are afraid of being phased out or fired outright, it makes change very difficult.

Sample of a Business that Changed as a Result of Scaling Back Operations

A mid-sized engineering firm was no longer as profitable as it had once been. As a result, the board decided it was time to scale back and focus only on a few key products that long-time customers still needed.

This meant that they no longer required some of the office space, equipment, and people they employed. They reshaped their operations to meet a smaller operational model.

In the aforementioned story, the company changed its operations in order to scale back on its supported products. Unfortunately, not many people were happy with this. Working with the team left behind was difficult to say the least. They simply did not want to change.

MERGERS AND ACQUISITIONS

Mergers and acquisitions (also known as M&A) are when two or more organizations either merge into one larger business entity; or one organization is acquired by the other. Either way, there are a lot of critical and time-sensitive decisions to be made that will impact how well the people within both organizations accept change.

How Do Mergers and Acquisitions Impact the Ability to Change?

Communication is the single-most important success factor in both mergers and acquisitions. Too soon, and people start getting anxious to get it over with. Too late, and people start feeling like something covert is going on.

Either way, the biggest concern that people have when it comes to mergers and acquisitions is job loss. Once that sets in, people go into full panic mode and start abandoning the organization as quickly as they can. What this means is that corporate tribal knowledge goes out the door with them to the competitors and those who are left most certainly are not going to accept change very easily.

Sample of a Business that Changed as a Result of M&A

An international insurance company acquired another insurance company in one of its many business deals. In order to merge the newly acquired entity into its operations as a division, the company decided to change the operating structure and merge the new products and services into its existing offerings.

This meant that people in the acquired company were at a high risk of job loss and the people adopting the new products into their departments had to restructure to accommodate and support the products.

In the preceding story, while many people did quit, many others were terminated. Many of those who were left were unwilling to share the knowledge they had with the people taking over the products and they did not want to give up on being the other company.

REFERENCES

1. http://www.businessinsider.com/how-lego-made-a-huge-turn around-2014-2.
2. Lego Group Annual Report 2015.
3. https://www.chargify.com/blog/6-companies-that-succeeded-by -changing-their-business-model/ http://www.foundersatwork.com/.
4. https://www.chargify.com/blog/6-companies-that-succeeded-by -changing-their-business-model/ http://www.foundersatwork.com/.

3

Interpreting the Language of Change

THE LANGUAGE OF CHANGE

The language of change, as illustrated in Figure 3.1, is almost as simple as it sounds.

It is the language that we use when talking about change to the people directing it, managing it, implementing it, and above all, to those impacted by it. In short, it is the words that we use in describing the changes and the expectations for adoption; and it's the words that people use in return, in order to express their feelings about those proposed changes. To that end, language is about mutual communication: both speaking and listening.

Habit #5

"Seek first to understand, then to be understood."[1]

How Does Language Impact the Ability to Change?

Language is everywhere and exists in many forms. However, that does not necessarily mean that it conveys the right message at the right time and to the right audience. What we say, or often do not say about pending change, plays an immense role in how well that change is accepted and adopted by those impacted most.

43

Figure 3.1   The language of change

In fact, language often exposes management's lack of understanding of their employees and customers, as well as their attitudes toward them. Therefore, the words that we choose to describe the changes can significantly impact the levels of engagement and adoption—either positively or negatively.

LISTEN: TO UNDERSTAND

Listening is the act of paying attention to and hearing what others are saying. It requires both seeing and hearing because the messages that people send can be verbal and nonverbal. Understanding others requires both elements to be present. The success of change initiatives depends on the consistency of messaging between both.

Active Listening

Active listening is a method for engaging in conversation with another person in order to hear and understand their concerns, considerations, and needs. To accomplish this, four basic steps are involved: listen, paraphrase, clarify, and respond (see Figure 3.2).

In Step 1—Listen—the change agent listens to the person speaking with many senses. They listen by watching people's body language, and they listen to their words as they speak. This enables them to gain a full understanding of what the person is saying and to gauge how genuine they are about what they are saying.

In Step 2—Paraphrase—it is important for the change agent to paraphrase what they believe the person has said and meant. To do this, the change agent repeats what has been said in their own words. To do this they use phrases such as: "I think I heard you say that…", "What I'm hearing you say is…", or "It sounds to me like you're asking for…".

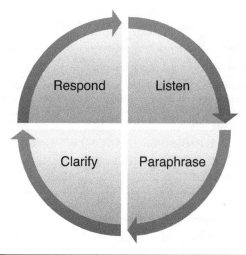

Figure 3.2   Active listening

Step 3—Clarify—is simple. It is where the change agent asks the person to validate their interpretation of what has been said. To do this, they ask a pointed question such as: "Is that correct?"; "Is that what you meant?"; or "Is that right?"

Active Listening Example: Part A

Bob's company is redesigning its interior workspaces to make the space more collaborative and people-friendly as part of a *Workplace of the Future* initiative. As a result, assigned cubicles will no longer be available and are being replaced with large open desks and seating areas.

Bob has approached Mary, the change consultant to talk about how unhappy this has made him.

Bob: "I don't know why we can't just keep things the same. It's worked for the past ten years that I've been here."

Mary: "What I think I'm hearing you say is that, you're upset that you have to change cubicles because you've had this one for ten years and don't see the need to change. Is that right?"

Finally, in Step 4—Respond—the change agent has an opportunity to respond. It is important to ensure that the response addresses what has already been said. This is not the time to go off on a tangent and spew the diatribe of why the project is important, and above all *do not* use this opportunity to regurgitate the project mission and vision verbatim.

Active Listening Example: Part B

Let's go back to hear the rest of the conversation between Bob and Mary.

Bob: "Yes. I just don't get it. It doesn't make sense to me."

Mary: "Well, Bob, if you recall last year, we held that week-long corporate session at the retreat and everyone said how much they loved working in the open with the teams. I recall you were one of the people suggesting a more collaborative workplace."

Bob: "Oh, I guess I thought it would be different than this."

Mary: "Well, let's talk about what you were expecting and see how that aligns to what we're doing."

Simply put, active listening helps people to feel as though their concerns have been both heard and addressed. This helps them to realize that they matter.

Active listening is important when it comes to discussions of change because despite being done in a professional setting, change is a deeply personal experience and it is one of the keys to making people feel important and empowered to make those changes. In the absence of all other knowledge about organizational change management, a change agent must be able to listen to those impacted by change because it is the single-most crucial building block of trust.

Active Listening Example: Part C

Imagine the conversation between Bob and Mary went something more like this:

Bob: "I don't know why we can't just keep things the same. It's worked for the past ten years that I've been here."

Mary: "Bob, we've been over this before. I'm not going to keep discussing it with you. If you don't get on board, I'm going to have to go to your supervisor and tell him what's going on."

Bob: "Go ahead. I've already gone to her and she doesn't listen either. It's like you guys just don't care about who is being hurt by these changes. All anyone cares about around here is the money we make for the company!"

Mary: "Bob, you know we initiated this project to make us more competitive in the marketplace by providing more opportunities for our team to collaborate."

Bob: "We're doing just fine as it is. We don't need this!"

In the above example, you can see that without active listening, this conversation took a much different tone, and Bob was left feeling defensive.

It is something so simple, yet very difficult to do unless it is practiced in conversations regularly.

How Does Active Listening Impact Change?

The impact that active listening has on change and change management is that it really provides a means of openly communicating true intentions and feelings behind things in ways that help others to fully understand what is meant. This makes change go much faster and smoother because people feel like they are a part of the changes rather than having it thrown at them when they are unprepared.

WHAT PEOPLE SAY AND WHY

Attitude Is Everything

An attitude is a way of thinking or feeling that is often reflected in behavior. In psychology, it is defined as *a relatively enduring organization of beliefs, feelings, and behavioral tendencies towards socially significant objects, groups, events, or symbols.*[2]

An attitude is usually the result of either experience or upbringing—or the culmination of both—and is a powerful influencing factor of behavior and, consequently, of the language utilized by people during organizational change efforts. Therefore, it is important to understand the concept of attitude in more detail.

Structure of Attitudes

Attitudes are most often described in terms of three components:[3]

- Affective:
    - How the other person, event, or object affects the feelings and emotional state of the central person.
    - For example: "I am terrified of heights."
- Behavioral:
    - How the attitude influences or affects the way that people behave.
    - For example: "I will drive for twenty miles just to avoid a high bridge."
- Cognitive:
    - What the central person knows, understands, and believes about the other person, event, or the object.

> ◻ For example: "I know bridges aren't safe because I watched one collapse just as I drove up to it."

These are illustrated in Figure 3.3.

Attitude Strength

Attitude strength is basically the degree to which an attitude is entrenched within a person's belief system. The amount of passion that a person has for a certain attitude is the result of the presence of certain conditions. These include:

- Attitudes that are the direct result of personal experience
- Level of proficiency and expertise in the subject matter
- When a favorable outcome is expected or anticipated
- Repetitive and consistent expression of the attitudes
- The stakes at risk (what they stand to win or lose)

Attitude strength is extremely important to understand when it comes to organizational change management because it can determine how likely it is that people will ultimately accept or reject change. Attitude, behavior, and language are inexorably linked. How people feel and perceive change is more often expressed in the subtle nuances of behavior and language; and it is the role of the change agent to ensure that attitude can be managed and reshaped throughout the transformation process.

In 1934, the LaPiere Study was conducted to understand the link between attitudes and behavior. What LaPiere concluded was that attitudes do not always predict behavior. However, his conclusion was based

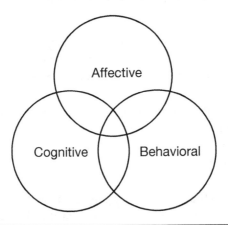

Figure 3.3   Attitude components

on a false assumption, and did not consider the differences between direct and indirect behavior. It is more accurate to say that attitudes do not predict behavior when confronted with a situation where the object of a negative attitude is present and in the presence of a person whose attitude is not known. This, in and of itself, decries the opposite of LaPiere's conclusion—that attitude *is* a predictor of behavior when all factors are considered.

This is precisely why people behave passive-aggressively. In fact, unless the transformation is well-managed, people will simply pretend to go along with the changes until the transformation initiative has ended and then they will go back to doing things the old way.

Why Is Attitude Important?

As change agents, knowing a person's attitude (as shown in Figure 3.4) helps us to predict their behavior when it comes to how they will react to and adopt change. In the absence of direct oppositional behavior, language is the single-best indicator for how people are feeling about and whether they will adopt the changes. What people say or do not say can tell others how they really feel.

Attitude

"Attitude is a little thing that makes a big difference." —Winston Churchill

---

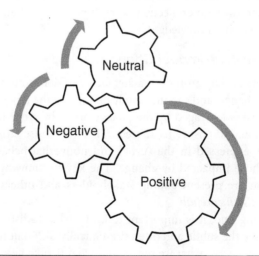

Figure 3.4   Attitude types

Self/Ego-expressive

The self-expression of attitudes is when people communicate how they feel and what they believe about a particular event, topic, or person. It is important because the attitudes expressed help to communicate who they are—and may make them feel good because they have asserted their identity.

Self-expression includes both verbal and nonverbal means of communication. This is why it is so critical to know how to interpret the language of change.

Adaptive

It is often said that people are *social animals* in that they deliberately seek out others who share their attitudes, and in turn, develop similar attitudes to those they like. The adaptive functions of attitudes help people to fit in with that social group. If a person expresses a similar attitude to that of the group—so-called *socially acceptable* attitudes—then others in that group will reward them with approval and social acceptance.

Ego-defensive

The ego-defensive function refers to having attitudes that protect a person's self-esteem or that justify actions that make them feel guilty. Positive attitudes toward themselves have a protective function (i.e., an ego-defensive role) in helping them to preserve their own self-image. The basic premise behind this is that attitudes help a person to mediate between their own inner needs (expression, defense) and the outside world (adaptive and knowledge).

How Does Attitude Impact Change?

Again, understanding attitude is a key factor in determining how people will react to change, and can ultimately determine the types of techniques and mechanisms leveraged by the change agent in order to manage the transformation and ensure its success. This is because attitude is everything and is expressed in the verbal and nonverbal behaviors demonstrated by those impacted by change. The key to knowing the kinds of attitudes that are present among stakeholders and others impacted by change is to *listen carefully*.

That being said, it is important to know what to listen for. In Table 3.1, we can see the subtle and often dramatically different interpretations between what we say, what we mean, and what people hear.

Table 3.1   The language of change

What We Say	What We Mean	What People Hear
We need to change how we work.	The current way of working no longer meets business needs.	The current way is broken.
		I know better.
		YOUR way doesn't work.
		YOUR way isn't good enough.
		You're going to have a new job.
		You may not have a job after this.
		You don't know what you're doing.
We need to be able to compete/get ahead of our competition.	The business shareholders want more.	You're not doing enough.
	We're losing customers and money.	We're losers.
		You're not good enough.
		You're not doing a good job.

Positive, Negative, and Neutral Attitudes

Essentially, there are three types of attitudes: positive, negative, and neutral. Each type is important to understand for a variety of reasons, such as how those having those attitudes will communicate, act, and help people through change.

Individuals with a positive attitude will pay attention to the good, rather than bad in people, objects, and events. They view mistakes or failures as opportunities; they learn from mistakes; and move forward in life.

Positive Attitude, Upbeat Language

"We can get through this together."
"What's next?"
"How do we come back together?"
"Let's do this."
"I'm game."

On the other hand, people with a negative attitude consistently ignore the good, while paying close attention to the bad in people, objects, and events; and they blame their failures on others. They are the most likely to complain about changes, instead of adapting to the changing environment.

Negative Attitude, Pejorative Language

"It'll never work."
"Management won't let us."
"We can't afford it."
"Even the instructor said this product was useless."
"It's all good in theory, but..."

Finally, people with neutral attitudes do not feel the need to change because they ignore the problem, leaving it for someone else to solve. They do not give enough importance to situations or events.

Neutral Attitude, Laissez-faire Language

"It doesn't matter to me."
"I'm not involved."
"Whatever."
"It's not important."
"So what?"

Does Language Evolve over the Transformation Effort?

Throughout the course of any transformation initiative, the language used by the people impacted by the changes will evolve one way or another. When change is well planned, managed, and executed, language shifts towards the positive from either neutral or negative. However, when change is not well-managed and executed, this language will shift toward the negative, regardless of where they started out.

This is because language evolves as people's attitudes shift and their confidence increases, but it also regresses as the opposite happens. Thinking in terms of the three types of attitudes, if a person changes from positive to negative, this will be demonstrated in their language. They may go from saying things such as, "We can do this, that obstacle is a challenge but it won't stop us," to "It can't be done, we've tried *everything*."

SPEAK: BEING UNDERSTOOD

What We Say, How, and Why

Delivering messages to the people impacted by change is just as important as hearing from them. Listening is only one-third of what it takes to earn trust, especially on a transformation initiative. It is also one part action and another part communicating with others (see Figure 3.5).

Figure 3.5   What is said versus what is heard

Considering the rest of this book is all about the actions that the change agent can leverage, this discussion is going to focus on the communication that those agents deliver back to those impacted by change.

Say What You Mean, and Say It Simply

What we say:
"The current operating structure will cease to exist."
What we mean:
"We are reorganizing the entire company."
What people hear:
"We're closing the company and you won't have a job."

What the change agent says about the transformation is crucial. It matters what is said in every single communiqué and conversation, because people are listening and trying to understand how change affects them and what the path to success looks like. It is important to write down and plan how the key messages are to be delivered and to practice them.

Saying nothing, on the other hand, is just as bad as saying the wrong thing. People will speculate and rumors will circulate. Be open and honest, and say things in a way that give people confidence—especially when there is a risk of job loss.

Never Say Nothing About Change

"Remember not only to say the right thing in the right place, but far more difficult still, to leave unsaid the wrong thing at the tempting moment." —Benjamin Franklin

---

Semantics

What Words Should Be Avoided at All Cost?

A double entendre is a particular way of wording a phrase that gives two meanings. Literally, it means double *understanding* or *hearing* in French. How change agents phrase conversations to others is important in order to prevent this from occurring. While active listening is a good way to mitigate this, being very precise in our language choices will also help. Table 3.2 outlines some examples of phrases and words that can be misinterpreted.

Table 3.2 Easily misinterpreted words

Words/phrases that may be misinterpreted or misunderstood	What others may think this means	Alternative words/phrases that are more supportive and informative
Should: "Everyone on the team **should**..."	It's a suggestion.	"Everyone on the team **will**..."
Could: "We **could** implement that idea in the next phase..."	It's going to happen.	"We will consider that suggestion for other phases and will decide on that when that time comes."
Must: "Customers **must** now..."	"I have no choice."	"Customers **can** now..."
Can't: "This idea **can't** be implemented."	This shuts people down from making suggestions.	"We need to do more research and analysis before we can determine how to handle this idea."
Won't: "The team **won't** be working on resolving that..."	Also shuts people down from making suggestions and focusing on the task at hand.	"The team has been asked to hold off resolving that until further research can be done to fully understand how to resolve that."
Acronyms and Jargon	Open to interpretation based on personal experience.	Use full names and phrases.
Try: "We are **trying** to implement..."	This is an attempt.	"We will implement..." "We are implementing..."
Closed-ended questions	Any question that can be answered with a *yes* or a *no*. This does not create dialogue that convinces people to adopt change.	Open-ended questions. Any question that elicits a full, detailed response.
"Please be advised..."	There is no room for conversation on this topic.	"We would like to let you know about..."

What Words Create Positive Vibes and Good Will?

Word Choice
"Words mean more than is set down on paper. It takes the human voice to infuse them with deeper meaning."[4] —Maya Angelou

By using the wrong words in the documentation and notifications for transformational initiatives, the door is left open for a *deeper meaning* to creep in. Unfortunately, in change management this means ambiguity and interpretation by the reader that translates into resistance. The objective of change messaging is to craft and deliver a concise message to the audience so that the buy-in is achieved. To this end, messages are more effective when the wording abides by some clear guidelines and avoids certain pitfalls.

Ambiguous Statements

Vague Executive Support

Vague executive support is where the sponsors of the transformation take a back seat and do not publicly communicate their support for the changes.

Ambiguous Scope

Ambiguous scope is when the messaging fails to communicate exactly what is changing including the boundaries of what is *not* changing.

Ambiguous Precedence

This is when the task order is unclear and the reader will be unable to determine the sequence of events and tasks in the change schedule. Table 3.3 shows examples of using ambiguous precedence.

Table 3.3 Ambiguous precedence

Poor Example	The purpose of this project is to transform the way we interact with our customers. This means we will be building new mobile applications and websites for customers, adding new computer terminals for them to use, hiring new staff, and we will be moving to a new office space to accommodate the larger customer service group.
Good Example	The purpose of this project is to transform the way we interact with our customers. This means we will be moving to a new office space, adding new computer terminals for customer service reps, building new mobile applications and websites for customers, and hiring new staff.

Ambiguous Adjectives

Adjectives are used to describe nouns and pronouns, and are used to embellish thoughts or to enhance reader understanding.

However, in change messaging, adjectives can confuse the reader because these terms are not specific enough to describe the changes, the solution, the expected participation, or even the problem. These problem adjectives are illustrated in Table 3.4; and Table 3.5 shows the usage of these terms in change messaging.

Ambiguous Adverbs

Adverbs are a common stumbling block for many business writers—and change agents are no exception. It is important to remember not to write the in the same way that people speak in conversation or for any other situation for that matter.

Writing change messaging is very different than writing an e-mail, a white paper, or a story. Adverbs are the parts of the descriptive makeup of a story that make it interesting. Change messaging must be specific, clear, and detailed—but it also needs to be concise and exact, so that people

Table 3.4   Ambiguous adjectives

all	any	appropriate
custom	efficient	every
few	frequent	improved
infrequent	intuitive	invalid
many	most	normal
ordinary	rare	same
seamless	several	similar
some	standard	the complete
the entire	transparent	typical
usual	valid	

Table 3.5   Statements using ambiguous adjectives

Poor Example	We will be offering a few training sessions for the entire staff to attend.
Good Example	To help support this project, we plan to run one training session per month for a year to ensure that you have a chance to find a session that meets your schedule. These sessions are mandatory for all staff who work in accounting.

know what to expect, where, when, and why. Typical adverbs that have no place in change messaging are shown in Table 3.6.

Table 3.7 provides an example of messages using ambiguous adverbs.

Ambiguous Synonyms

Using a vague term to replace a specific name is an ambiguous synonym. Wording and phrases to watch for are shown in Table 3.8.

Table 3.6   Ambiguous adverbs

accordingly	almost	approximately
by and large	commonly	customarily
efficiently	frequently	generally
hardly ever	in general	infrequently
intuitively	just about	more often than not
more or less	mostly	nearly
normally	not quite	often
on the odd occasion	ordinarily	rarely
roughly	seamlessly	seldom
similarly	sometime	somewhat
transparently	typically	usually
virtually		

Table 3.7   Statements using ambiguous adverbs

Poor Example	Please dress accordingly for the upcoming event.
Good Example	This event is an outdoor sporting event. You will be running and getting dirty. Please dress in sports wear.

Table 3.8   Ambiguous synonyms

the application	the component	the data
the database	the field	the file
the frame	the information	the message
the module	the page	the rule
the screen	the status	the system
the table	the value	the window

A good rule of thumb to follow is that if there is a need to use the name more than once, the message is vague and needs to be reworded. Examples of messages using ambiguous synonyms are shown in Table 3.9.

Ambiguous Verbs

Verbs can be a significant source of confusion in messaging because there is a tendency to write passively so as not to force people to change. The truth is that people are expecting clear and decisive messaging that gives directions about what is changing and why, as well as who is impacted; but they also expect to see directions for how to contribute—what they should do and when. Table 3.10 identifies specific terms to avoid, and Table 3.11 shows examples of messages using these terms.

Table 3.9   Statements using ambiguous synonyms

Poor Example	The components of the system will be implemented on the weekend. Please ensure your computers are left on over the weekend.
Good Example	The XYZ software application will be installed on Saturday after 12 AM. Please ensure that you leave your computers logged out, but turned on.

Table 3.10   Examples of ambiguous verbs

adjust	alter	amend
calculate	change	compare
compute	convert	create
customize	derive	determine
edit	enable	improve
indicate	manipulate	match
maximize	may	minimize
might	modify	optimize
perform	process	produce
provide	support	update
validate	verify	

Table 3.11   Examples of statements using ambiguous verbs

Poor Example	We will be amending the training schedule to minimize disruption during the workweek.
Good Example	We will be making changes to the training schedule. Currently, training is offered at peak customer call times. To support your ability to continue to provide excellent customer service, we are moving training to the weekends. A full schedule will be e-mailed to you within one week.

Built-in Assumptions

Functional/Environmental Knowledge

There is an element of assumed knowledge for the audience of transformational messaging. Change agents and project teams alike often assume that line-employees, vendors, partners, and customers all have pertinent knowledge regarding the systems, processes, and policies that are impacted by change. This means the authors (in most cases the change agents) are building-in a level of tribal knowledge that is not readily translated across the business.

Use of Jargon

Techies and developers tend to use a lot of slang and jargon. Jargon is abbreviated terms such as *ASAP* instead of *as soon as possible*. If abbreviated jargon terms are used repeatedly throughout the change messaging rather than writing out a full phrase, it gets confusing.

Directive

It is important to ensure that change messaging clearly states the expected actions to be taken by those impacted by change. This messaging is a set of directions as much as it is to inform people about impending change. The change agent is not making personal requests to do favors, and they must be worded as such.

Words like *should* and *may* have no place in change messaging because they indicate a moral imperative and, if used, that the action could be ignored by those who must perform it. Table 3.12 illustrates examples of good and not-so-good messages.

Implicit Cases

Another common mistake that people tend to make in writing is that they imply other attributes and inclusive cases. While ambiguous implicit cases are quite acceptable in colloquial writing, change messaging must be precise and must therefore be written in a way that leaves no room for misinterpretation. Table 3.13 identifies terms which create implicit ambiguity, and Table 3.14 presents examples of ambiguous messages using implicit cases.

Negation

Scope of Negation

Scope of negation refers to specific boundaries for a negative item. It is important that it is explicit which item in change messaging is being

Table 3.12 Statements using ambiguous directives

Poor Message	All customer service representatives should attend one of the informational sessions.
Good Message	Mandatory information sessions have been set up every Wednesday for the next month. All customer service representatives are to sign up for and attend one of these informational sessions.

Table 3.13 Ambiguous implicit cases

also	although	as well
besides	but	even though
for all other	furthermore	however
in addition to	likewise	moreover
notwithstanding	on the other hand	otherwise
still	though	unless
whereas	yet	as required
as necessary		

Table 3.14 Ambiguous messaging using implicit cases

Poor Example	Attending the training is mandatory for all employees unless they work during the weekend.
Good Example	Attending the training session this weekend (March 15, 2017) is mandatory. Only those working on March 15th will be scheduled on a day in April. This date will be set next week and you will receive an e-mail with the date and time.

negated so that people completely understand what is and is not being done. Table 3.15 provides both good and poor examples of the scope of negation.

Unnecessary Negation

Unnecessary negation simply refers to situations where the change agent is using more simplistic language. If the item can be written without negation to be clearer, then it must be written without the negative. Table 3.16 provides examples.

Double Negation

It is common for people to use double negatives in colloquial (conversational) language. This is a bad habit which must be avoided in change messaging. Table 3.17 provides examples of good and poor examples of double negation.

Table 3.15 Statements using ambiguous scope of negation

Poor Example	If employees are intending to attend the weekend retreat, and they have not already signed up, they not only need to sign up online but also send an e-mail to the administrator to ensure there is a space available for them.
Good Example	All employees wishing to attend the weekend retreat must sign up online and send an e-mail to the administrator to verify that space is available.

Table 3.16 Statements using unnecessary negation

Poor Example	If employees are not already scheduled to work, you need not only show up for work as per usual, but also attend the training workshop afterwards.
Good Example	All employees who are on a day off must come in to work on the day of the workshop and work a regular shift before attending the workshop in the evening.

Table 3.17 Statements using double negation

Poor Example	Employees cannot choose not to attend the meeting.
Good Example	This meeting is mandatory for all company employees.

Scope of Action

It is important to ensure that change messaging completely and fully defines the changes being made and what those who are impacted must do to prepare and adopt those changes. Table 3.18 illustrates this concept.

Time Reference Ambiguity

Time reference ambiguity occurs when the change messaging makes vague references to time. All change messaging must be clear about time-lines and known schedules so that people can prepare both mentally and emotionally for the changes. Table 3.19 identifies specific terms to watch for and clarify.

Messaging Mediums

One of the things that a change agent will undoubtedly be called upon to do is to communicate and liaison with those impacted by change on a one-to-one personal basis. E-mail is likely to be a big part of this, therefore it is important to know not just what to write, but how to write

Table 3.18 Statements using scope of action

Poor Example	All systems in the accounting area will be migrated over the weekend. Some training session are available and as of Monday we will be fully using the new software.
Good Example	All accounting computer systems will be upgraded with new software this weekend (March 15, 2017). Training sessions have been scheduled for Friday, March 14, and all accounting staff are required to attend. Before you leave on Friday, March 14, please log off of your computer and leave it on for this upgrade to occur.

Table 3.19 Time references

after	annually	at a given time
at the appropriate time	bimonthly	biweekly
daily	every other month	every other week
fast	in a while	later
monthly	quarterly	quickly
soon	twice a month	twice a year
weekly	yearly	next
former	latter	previous
quickly	slowly	

e-mails that get opened, read, and actioned. The following guide will provide a framework for writing effective e-mails.

Subject Line

What is it? *Tag line.*
   What does it include?

- Statement of the level of action expected from the reader:
    - For Your Information
    - Response Requested
    - Response Required
    - Urgent Response Required
    - Action Required
    - Urgent Action Required
    - Assistance Requested
    - Urgent Assistance Requested
    - Information Requested
- Subject: Clearly state the subject you wish them to address

---

Subject Line Example

Action Required—Mandatory Technical Training Course Enrollment

---

Body of E-mail

What is it? *Main message.*
   What does it include?

- Salutation
- Set tone: Thank them or tell them something you appreciate about them
- Purpose: Tell them *exactly* why you are contacting them—in one sentence
- Action requested: Tell them *what* action you expect from *whom* and by *when*
    - Request additions/contributions to the agenda
    - Contributions to discussions—cite topics
    - Prepared topics and discussion leadership
    - Brainstorming
- Detail the type of preparation expected
    - Send documents that need to be read before the meeting with enough advance notice to read them

       ◻ Samples of documents and artifacts they will contribute

       ◻ Demonstrations and exercises to be done

       ◻ Include agendas for meetings

Body Example

Good Morning Dale,

Thank you for taking the time to attend our recent technical training session and offering to participate in a focus group for the upcoming transformation effort.

I am contacting you today to ask you to attend the first of those focus group sessions being held this Friday at 2 PM. Please RSVP by tomorrow at noon.

This first session will discuss the following items:

1. Making the changes successful
2. Getting more support and participation from teams
3. Team building activities we can conduct

I would also like to ask you to bring to this session any ideas and examples of fun team building activities you have attended in the past.

Close

- Thank them for their time

Closing Example

Thank you! I look forward to seeing you at the session.

Consultation

Consultation is a framework for objective discussion aimed at finding the truth and sharing a diversity of perspectives in a safe manner that is geared toward creating greater understanding of the people involved. It is active listening in a group setting. It leverages a set of guidelines for the discussion that enable every participant to contribute in meaningful ways that help the group to arrive at consensus.

Consultation is important when it comes to change because key stakeholders are most often the ones who will relay information outward. While they won't be the sole source for information about the transformation initiative, they most certainly will be interpreting what they hear and feel into conversations among their own teams.

By leveraging consultation in discussions with these key stakeholders, the change agent is effectively consulting with people who may not have the opportunity to attend as a direct participant. When stakehold-ers—as representatives for their teams—are treated with respect and con-sulted through this process, it effectively gives a voice to everyone on the team, even when they are not in the room.

How Does this Impact Change?

Change becomes easier when more people feel as though they are included in the process and especially in the decisions made throughout that process. The vast majority of resistance to change is because of fear that results from a feeling of a lack of control. Remember, *people support what they helped to build.*

Conducting Effective Consultation Sessions

Set-up:

- Set an agenda and e-mail with the meeting request
- Request additions/contributions to the agenda
- Determine the setting based on the type of meeting you are having
    - People will only participate if they are listening
    - People will only listen if they are comfortable
- Determine if a break will be necessary
    - Meetings over 2 hours
- Detail the type of participation that you expect
    - Contributions to discussions—cite topics
    - Prepared topics and discussion leadership
    - Brainstorming
- Detail the type of preparation expected
    - Send documents that need to be read before the meeting with enough advance notice to read them
    - Samples of documents and artifacts they will contribute
    - Demonstrations and exercises to be done

At the session:

- Review the agenda
- Thank participants for their time
- Ask if anyone needs to get a beverage
- Start main topic first
- Before each discussion, let people know the level of participa-tion during the talk

- ◻ Questions during the talk
- ◻ Save questions and comments until after
- Ensure time for discussion between topics
- Monitor time and arrange for another time if there is more to discuss
- Manage participants to ensure everyone contributes
  - ◻ Directly ask those not speaking if they have anything to add
- Close each topic before moving into the next

Close the meeting:

- Review any action items identified during the meeting
  - ◻ Cite who is to perform the action and by when
- Set the agenda of missed items for the next meeting
- Ask for additional items to be added to the next agenda
- Set a date and time for the next meeting if possible
- Thank participants

REFERENCES

1. Covey, Stephen R.; *The 7 Habits of Highly Effective People*; https://www.stephencovey.com/7habits/7habits-habit5.php.
2. Hogg, M. and Vaughan, G. (2005). *Social Psychology (4th edition)*. London: Prentice-Hall.
3. http://www.simplypsychology.org/attitudes.html.
4. Angelou, Maya; http://www.brainyquote.com/quotes/quotes/m/mayaangelo140532.html.

The Process of Change

The process of change as described in Prosci's ADKAR model is that people change when they first have *awareness* that change needs to happen; a *desire* for the changes to occur; *knowledge* about how to make the changes that are necessary; the *ability* to make those same changes; and the external *reinforcement* to do so. All of this would suggest that change is an inherently internal individual process, and that successful change management efforts must seek to leverage this process in order to effect changes within groups.

However, Behavior Modification Theory[1] suggests that change can also be completely external to the individual because without any awareness, desire, and knowledge on the part of the individual, changes can be affected through positive and negative reinforcement. This is especially important when people cannot or will not be able to grasp the context or the rationale for the changes.

The process of change is both internal and external. The objective of change management efforts is to leverage a balance of both internal and external techniques throughout an applied methodology, in order to create lasting and sustainable changes within a given organization.

Getting Lost

When I was younger, if I was driving somewhere new and got lost, I would start to feel the panic set in because I was feeling a lack of control. The radio went off, no one was allowed to speak to me, and I would sometimes cry and yell. While this is admittedly not a very flattering self-portrait, it is important to know that many people feel the same when changes are announced.

The way I have learned to work past it is to stop the car, pull out the map or the GPS, take some deep breaths, and take a break from driving.

What we can learn about change management from this is that people need to have some sense of control over what happens around them (change cannot happen *to* them.) Their sensory receptors will often go into overload and they will react. These are indicators that can expose the levels of stress that people are feeling as a result of the changes. Why is this important to know? Quite frankly, it's important to know so that change efforts can be successfully managed by monitoring these kinds of stress indicators.

Why don't organizations employ only internal *or* external change processes in order to transform? Because if they only leveraged internal individual change processes, they would end up managing to the exception (like some Monty Python skit about how all the decisions of the office are ratified at a special bi-weekly meeting); and if they only leveraged external change processes, they would end up with a high attrition rate in a dictatorship.

That being said, it is important that organizations take a balanced approach to managing change efforts in order to create the most holistic transformation that is organization-specific (unique to each organization). To do that, it is imperative to know and understand the entire internal process of change (how individuals change themselves) and the external process of change (how organizations change individuals).

INTERNAL CHANGE: HOW INDIVIDUALS CHANGE THEMSELVES

People at an individual level implement change into their own lives through a simple pattern: They must acknowledge and accept that change is necessary; be willing to accept the proposed changes as the right course of action; decide that they are open to changing; have confidence that they can change; and be able to move from inconsistent to consistent application or demonstration of the changes. This pattern is depicted in Figure 4.1.

Acknowledge and Accept Change

Before people can even begin to move through the process of changing, they must first, and foremost, acknowledge to themselves that change is, in fact, necessary. Acknowledgment is merely the admission that change is the only logical course of action.

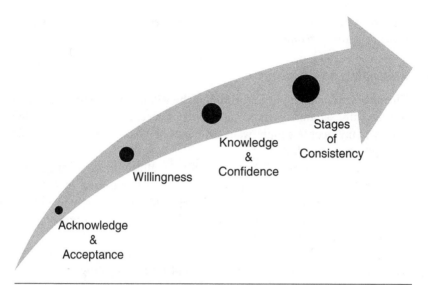

Figure 4.1   Individual pattern of change

Weight Loss Acknowledgment

Before a person can embark on any kind of weight loss initiative, they must first admit or acknowledge that they need to lose weight.

Acceptance, on the other hand, is a two-edged sword. Once the person has acknowledged that change is necessary, they must also accept that they will attempt to make the changes, and they must also accept the proposed changes themselves.

Weight Loss Acceptance

Once the person has acknowledged that they need to lose weight, they must accept that they need to lose it. This is probably the hardest part because they will likely know that they need to, but still cannot accept it. The good news is, however, that once they are over the first hurdle of acceptance, they only have to accept the means for losing weight. Those means could be dieting, exercise, or even surgery.

Like weight loss, all change is hard because once people have made the decision to change by both acknowledging and accepting change is necessary, and that this project is the best means of making those changes, they just wish it was over. There is never a magic switch that can be flipped—change requires consistent, concerted efforts of the group in order to come to fruition.

Be Willing to Change

The next step that people go through is the willingness to change. This is not the same as merely acknowledging and accepting change. As stated above, many people simply want to make the changes without actually having to go through them.

The truth is—some of them are still not really willing to change. This is evident because if there was another way to get around making the changes, or they could simply not change at all, many people would not make them (as an aside, this is exactly *why* the weight loss industry is a multibillion dollar industry).

People actually bounce back and forth like a pendulum between acknowledging change must occur, accepting it, accepting the specific course of action, and willingness to change. Why? *Fear.* As discussed in Chapter 1, fear is one of the biggest barriers—if not *the* biggest barrier—to change.

It is the role of change agents to create an environment and a strategy that takes this into consideration and allows people to go through this until they are comfortable. They only thing that makes people feel comfortable is having the knowledge and confidence in that knowledge to perform the new role or to tackle the changes.

Have Knowledge and Confidence

Knowledge and confidence are simple. People do not need to know and understand every single detail of the change that is coming, but they do need to know the basics, such as: how change will occur (what is the orderly process for evacuating the building, so to speak), what their specific role will be, how the changes will impact them, and of course, what the changes will be.

Once they have this knowledge, they need time and opportunities to go over the plan and learn it so that they can gain the level of confidence required for them to make the proverbial leap of faith.

> **Enterprise-wide Transformation Effort**
>
> Recently, a mid-sized multinational insurance company decided that they needed to undertake an enterprise-wide transformation effort in order to change the way that they conducted business. They wanted to avoid falling so far behind the competition that they would eventually go out of business.
>
> One particular senior executive decided that she did not agree with the transformation because she had neither acknowledged its necessity nor accepted the specific plans to change. She was simply unwilling to change.
>
> This particular executive did not have the knowledge and confidence to enable the changes in her division, simply because she had not had the time to go through the process.

Move from Inconsistent to Consistent Application of the Changes

The important thing for change practitioners is to understand how this simple pattern of individuals changing themselves influences and informs the process of organizational change. Once people have changed, the techniques can be applied to support, enable, and engage them in changes at the individual level.

Organizations don't change, people do. The truth is that organizations are transformed as the people within them change and adopt new behaviors and ways of doing things. As a change agent, it's important to remember and understand that.

INTERNALIZATION: HOW PEOPLE PREPARE

Many people prepare for change by first internalizing it and moving through the five stages of grief before they can actually accept change. Change must be internalized before people can actually adopt it and move on.

5 Stages of Grief

As depicted in Figure 4.2, according to Kübler-Ross,[2] people go through emotional cycles of denial, anger, bargaining, depression, and finally, acceptance when they are confronted with grief. Consider this: people deal with these very same cycles when confronted with change—especially when that change is dramatic.

Looking back to the start of this chapter, recall that the discussion was focused on people having to acknowledge change before they can

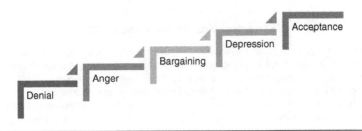

Figure 4.2   Kübler-Ross—5 Stages of Grief

accept it. The opposite of acknowledging the need for change is denial. So, until people acknowledge the need for change we can say that they are in a state of denial—or ignorance.

Next is anger. Where denial says, "I don't need to change"; anger says, "I don't need to change, I'm fine like this." When people are angry, they cannot change. They are angry because they have suddenly realized that the change is needed, but they're not happy about it.

Here is the problem: when people are angry they do not think clearly or rationally. This is the worst time for making decisions, and getting commitment and consensus on proposed changes is not typically possible during this time.

After anger, comes bargaining. It looks and sounds a lot like, "I'll do this instead," or "What if we make this change and not that one?" In bargaining, people try to negotiate not only an alternate route to the changed state, but they also try to negotiate with themselves about what they'll do to convince themselves to change.

Depression is the stage where people are most likely to experience sadness and hopelessness. This impacts the ability to change in a big way—if it is allowed to continue for any length of time—because people will not participate in the transformation if they feel as though they are not important and their efforts are useless. Some of the most common things said are: "What's the point, no one else will bother?" or "My vote doesn't count anyway."

Acceptance is the final stage, where people accept that change is not only going to happen, but that the proposed solution should go forward. However, acceptance is not—and should never be—mistaken for buy-in or engagement. Remember, buy-in means believe in, engagement means involvement, and acceptance means passivity.

Visualizing Success

Before people can feel as though they going to be successful, they need to first know what success looks like and they need to be able to see themselves as being successful. Far too often, people know what success looks like but simply don't see themselves as being successful. It is the role of the change agent to ensure that everyone can visualize being successful with the changes. This is the whole reason for creating a unifying project vision and mission statement; and for leveraging specific tools that not only allow people to know what success looks like, but to also see themselves as successful.

Jetty Jumping

It takes a lot of courage to take a leap of faith when inside you are petrified of the potential outcomes. Jetty jumping is a euphemism for taking that leap.

Overcoming Fear

When I was fourteen I went to a military summer camp. Me—terrified of dark water, crippled by fear of dark open spaces, and petrified by a fear of heights. What could possibly go wrong?

Well, having to jump off of a 35-foot-high jetty into the ocean, actually. So, there I was (there is video footage to prove this by the way), knees shaking, pulse racing, chest feeling like it was closing in, standing at the top of the jetty looking for a way out—any *other* way out.

After 30 minutes the rest of my division had all jumped and given up and gone back to the barracks. Great. Now I had to jump *alone* into that dark watery abyss below. I should have jumped while everyone else was still there (in case something really did want to eat me). The only thing on my side was the tide.

After 56 long, grueling minutes of swimming off the lower flotillas, and small practice jumps off of a shorter ledge, I finally mustered the courage to force myself off the big ledge. Swimming as fast as my arms and legs could carry me, I made it back to the flotilla and was allowed to leave.

Despite still trembling with fear, I was proud of myself. But in that moment, I also realized that most of fears were in my head. That summer, I forced myself to jump another time, and the following year I jumped four times and also crushed walking across the telephone pole on the confidence course.

I'm still afraid of things. But I work to overcome them so that they never stand in my way again.

Jetty jumping represents the internal battle being faced on every single project or change initiative—getting people past the fear that can be larger than life, no matter how irrational it can be. It's a battle against fear of the unknown and fear of bad things happening as a result. It takes a lot of communication and coaching to help people find ways of diminishing their fears—just long enough to make the leap.

HOW PEOPLE CHANGE

Imagine that you've suddenly gone blind. Not gradually over time, but you wake up one morning and cannot see anything—it's all dark, no light, no blurred shapes, nothing but blackness. How would you feel? How would you navigate around your own room? After all, it's nothing you've prepared for.

Sudden change can have that effect on people. When they wake up, come to work, and discover that something has changed—they find themselves fumbling in the dark. They bump into things, and get frustrated, and react in negative ways; and all because someone moved the chairs while they were out. Okay, moving chairs is a minor exaggeration, but what is important to note here is that change needs preparation. It needs early warning, communication, and managing expectations.

Change is deeply personal. It hits how people feel about work, about others around them, and even about themselves. While change initiatives are often highly prescriptive, their impacts are anything but. It is important to remember that.

People feel change before it ever even happens. Because of this, change management efforts must target the emotions of the people who are impacted by change, and they must do it quickly.

Overcoming Learned Helplessness

One of the things that people experience during the initial stages of change is the fear that comes along with learned helplessness. This is because they have not been there or done that—or worse, they have tried it and failed.

The most famous story of learned helplessness is the story of the elephant discussed in Chapter 1. Even though the elephant is quite large and can easily move the small bush, he doesn't *think* that he can, because he tried when he was small and was unable to do it, so he eventually stopped trying.

Learned helplessness is when people give up trying to change because they feel like they cannot do it. Make no mistake; this is fear masquerading as self-preservation. The truth is that by refusing to change, they are effectively making themselves obsolete.

How Do We Overcome Learned Helplessness?

Just like jetty jumping, it takes a lot of coaching to overcome learned helplessness. However, it also takes a lot of guidance for people to realize how they can change and still be successful. The most powerful tool for overcoming learned helplessness is *quick wins* or pilot projects.

Make no mistake, the role of change management is to find these people and to help them learn how to be successful once the changes have been implemented. It is not the role of change management to help the keeners (the active participants) in the crowd, who already believe in the changes, to adopt them. That would be like asking people already on the baseball team if they want to play.

Getting (and Feeling) Support

Finally, it takes feeling a connection to others on the team that enables people to make the leap of faith. Look at it this way, would you jump out of an airplane knowing that the person who packed your parachute is in your will and after your money? Probably not. But you would if you trusted the person who packed it.

You would if you trusted them with, well, say, your life. Why? Because you know they have your best interests at heart, and are not out to get you. But trusting you is only a part of the equation. They have to feel supported during the process of change. In other words, they have to believe in the parachute as well. They have to know that it will support them all the way down.

Support is the encouragement, backing, and assistance to change— especially when change is hard and uncomfortable. To provide support, change managers must be prepared to connect with the stakeholders and people impacted by change. They need to listen to them and encourage them to share their concerns, address them, and help them find success.

How Does Support Impact Ability to Change?

Having support is literally the difference between changing and not changing. I once heard a change manager write off a large segment of the company's population during a massive enterprise-wide initiative. She claimed that she knew not everyone would get behind the changes being made. Worse yet, she didn't try to create strategies to help them or even to reach out and connect with them.

Is it time consuming, yes; but when the person holding out is a vice-president in the company, those responsible for change management need to take that seriously and address it. These people need support. They need you.

HOW PEOPLE INCORPORATE CHANGE INTO THEIR NEW ROLE

Try It, Buy It, Use It

The best way to engage stakeholders and those that will be impacted by change is to provide time and opportunities for them to try the changes in a safe place, before they are implemented. It is the same simple concept as it is for making any other significant purchase, with one difference: purchasing a home or a vehicle is a personal decision, not a corporate one.

That means that people are not automatically in love with the product before the purchase, because the purchase is made by evaluating alternatives in a very impersonal way. No one sells software or consulting services by asking if you can see yourself having a barbecue in the backyard or driving along the coastal highway with the top down.

Sharing Success

It is only after people have had the chance to try the product out and discover how well it supports them in doing their work, that they will share their newfound enthusiasm with their colleagues. This socialization is absolutely critical in the process of enabling and managing change.

This type of socialization enables the change to become a grassroots movement that dramatically reduces change management efforts and increases the adoption rate of those impacted by the change.

Mentoring Others

Once people *on the ground* begin to share their enthusiasm and successes with the new products and the new way of working, it is natural for them to start mentoring others around them as a part of this sharing process. In other words, in the process of sharing our excitement and spreading the good news about this new *shiny* tool or process being implemented, we feel compelled to help others to become as excited and well versed in it as we ourselves have become.

It is a part of the natural progression because of our need to feel connected to one another. We naturally want everyone around us to be as happy and satisfied as we are. So we help them get there by supporting them and teaching them what we know.

Run, Don't Walk

This information about the *natural progression* of change is important; and when the agents of change are aware of that, they can capitalize on it, plan for it, support it, and even sponsor it—but it has to be done early. There is a narrow window of time when you can capitalize on the trust of your audience to be able to ensure that they are quick adopters, start sharing their successes, and mentor others.

EXTERNAL CHANGE: HOW ORGANIZATIONS CHANGE INDIVIDUALS

The external elements of change are those that come from the organization when change is about to be imposed on the individuals within. In truth, change is not really being *imposed* so much as being *proposed*. Again, there must be a balance between the individual and the organization when it comes to the power and control over the changes being made. It is important to understand that there are times when change is initiated by one body or the other—the individual or the organization.

That being said, all change efforts require the support and engagement of *both* in order to be successful. The accountabilities and responsibilities of the organization during the effort (transformation initiative) are to communicate proposed changes and rationale (drivers and objectives behind change) to all those directly and indirectly impacted; to set and manage expectations; to engage people throughout the process; to provide opportunities for knowledge transfer and building individual confidence; to provide ongoing support to those impacted; to provide positive reinforcement and a reliable governance structure; and to demonstrate the effectiveness of the change efforts.

COMMUNICATE PROPOSED CHANGES AND RATIONALE

Here is where I tell you that change management isn't necessarily associated with project responsibility and accountability. It is, in fact, an effort that supersedes the project and is more about managing organizational transparency than it is about controlling how well people accept changes that are imposed as a result of an individual project.

This is true because change happens regardless of individual project efforts. How people accept and become involved in the change is a product of how they feel about the organization. But it's also true because when *communication* about the need for change occurs (creating awareness), it must happen before individual projects are initiated. It must occur at the strategic planning level (as illustrated in Figure 4.3).

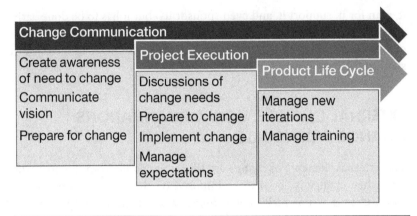

Figure 4.3 Change communication layers

That is not to say that change management itself is not an activity that should be utilized on individual projects, it is however, to say that change management activities must be both tactical and strategic in nature in order to adopt a culture of change, as discussed in Chapter 12.

Successful change—organizational transformation, if you will—begins with open, honest, and appropriate communication from the executives. Depending on the drivers for change, this communication can be either an acknowledgment that validates the suggestions or concerns of the individuals that work or interact with the organization; or it can be notification that the organization must change as a result of factors that individuals may or may not be aware of.

SETTING AND MANAGING EXPECTATIONS

One of the characteristics of a successful change effort—and arguably, the project itself—is the ability of the team to set and manage expectations. Many people will come in with their own individual expectations. The change manager must work to align those expectations so that everyone is on the same page. Communication is one of the most important methods for achieving this alignment.

The crucial factors to setting and managing expectations are having transparent discussions to present the transformation mission, open communication lines, demonstrate consistency of messaging among the executive, show cohesiveness of the organization, and leverage communication architecture. Managing expectations is a sure-fire way to enable and support engagement.

Having Transparent Discussions

Change management is not like playing poker—all of the cards must be on the table. There is no one-upmanship because the trust of the people impacted by change is at stake. If these people do not trust their executives or the team implementing change, they will not adopt the changes regardless of how much they are bribed, threatened, or even how great the solution is for them. It could be plated in solid gold and ready to sit on their desk, and they will resist taking it.

Because of this, it is important to have frank and transparent discussions to present the mission of the transformation initiative as early as possible. This is an important element in setting and managing change by managing expectations because people have an easier time changing when they know what is coming and have time to digest and accept the changes.

Open Communication Lines

The very art of managing expectations dictates that the lines of communication remain open. The key is that they will only remain open if there is trust between the parties and discussion occurs when changes to the established expectations occur.

Demonstrate Consistency of Messaging

In addition to maintaining open and honest communication, it is absolutely imperative that EVERY executive in the company, and all of those among the vendor partners, deliver the same message to those expected to adopt change. By delivering consistent messaging from all of the executives and the vendors, it will be easier to achieve and maintain people's trust, and trust is the single-most important reason for managing people's expectations.

Demonstrate Cohesiveness of the Organization

Transparency, open communication, and consistency of messaging all point to and demonstrate cohesiveness of the organization. When there are fractures in an organization, people not only find it hard to trust, but they also do not follow. This means that they will not adopt the changes being implemented, or will do so only half-heartedly—as if they are playing along. The changes will not stick and many of the people will revert back to their old ways of doing things.

Leveraging Communication Architecture

Finally, in order to accomplish all of these magnanimous tasks, it will be important to leverage communication architecture designed to execute and govern all aspects of the transformation initiative. Communication architecture, as described in Chapter 8, is the planning and infrastructure which enables consistency, openness, and transparency of the messaging at the appropriate times, to the appropriate audiences.

ENGAGE PEOPLE IN CHANGE

The key to any successful change or transformation effort is to engage people in the process. While this is often referred to as *buy-in*, the term engagement is more appropriate because it refers to the involvement, participation, and contribution of those people who are both directly and indirectly impacted by the changes being implemented.

Making Sustainable Changes

People support what they helped to build.

When people are engaged, it can be said that they are actively connected to what is going on around them. Does it mean that they are involved in every aspect and decision to be made? No, it means that they are contributing their expertise to the changes being made—where and when those changes require their expertise.

How do you get five or ten thousand people to participate and contribute to the changes being made? Committees, surveys, workshops, skunk works, hackathons, and pilot projects are a good place to start. One of the critical factors in engagement is for people to feel important and heard. So, while you do not have to incorporate *every* single suggestion (people don't expect you to), you have to make people feel as though they have been heard by acknowledging them.

In a large change initiative, if a survey is sent out, ensure that when you report the results, you cite all the great ideas that came forward; how difficult it was to choose the most important ones to implement; and give examples of some of the ones that were not used. This goes a long way to saying we heard you. However, the one thing that can and does go wrong with surveys and other feedback mechanisms is that organizations ask for suggestions, and then blatantly ignore them all. That's called lip-service;

when you report how great the suggestions were and yet don't use any of them.

The truth is that engagement is about communication. It's about feeling heard, and feeling heard makes people feel important. The more important they feel to the process, the more involved they will be. It's just that simple.

Perhaps you're sitting there thinking that people often do not take the opportunities in front of them. That is true, opportunity does not equal participation. There are two significant factors why people don't take advantage of opportunities: timing and feeling important.

Stop and think about it—was enough time given to allow people to participate in the training? Hint: if it was offered three times over the course of a single month for 100 employees or more, the answer is no. Timing of opportunities is just as crucial as feeling important to the process of change.

PROVIDE OPPORTUNITIES FOR KNOWLEDGE TRANSFER AND BUILDING CONFIDENCE

As we've just discussed, the timing of opportunities is as imperative as making people feel important, when it comes to ensuring that they will capitalize on the opportunity to contribute or participate. That being said, one of the critical factors in engagement is helping people to feel confident. As discussed earlier in this chapter, confidence is a product of knowing where the process leads, what success looks like, and trusting in leadership to be supportive during the trial and error portion of the implementation phase.

The Nonparticipant

Everyone has been in at least one meeting where one or two people sat there and contributed nothing, even when prompted or asked. This person might as well have been absent or asleep (if they weren't actually sleeping), for all they participated in the meeting.

This person is called the nonparticipant and will usually opt to pass and not provide their insight, opinion, or other feedback. The reason is that this person is checked-out. They are so disengaged that they really do not care about contributing to the group.

The primary reason for this disengagement is that they feel unimportant. What they have to say, what they think, and what they know is not important to the process.

Ultimately, people will only change when they are confident. The truth is that getting there is a journey unto itself.

Quick Wins and Pilot Projects

Quick wins and pilot projects are often considered to be one and the same; however, this is not the case. While a pilot project can be considered a quick win, it is not always the case, and a quick win does not always take the form of a pilot project.

As we will further discuss in Chapter 10, quick wins are any activity that can be leveraged within a short period of time (over the course of less than one month) to affect a shift in culture, thinking, and emotional response to the proposed changes. Typically, quick wins are very short efforts of one week or less that are run over the course of the project in short spurts.

Example of a Quick Win

An example of a quick win is using an activity to help people feel comfortable with the new processes and products. One company leveraged a cooking class to teach new business analysts how to collect requirements. The analysts had almost no previous knowledge of technology projects and certainly had no experience in eliciting requirements from stakeholders. To say that they were resistant is an understatement.

The class was designed in such a way that participants had to organize themselves into teams, then had to elicit requirements for a meal from a stakeholder. After eliciting the requirements, the teams made the lunch according to the stakeholder's specifications and ate the meal.

This activity was a quick win because it enabled the new analysts to leverage skills they would need to perform their new roles and showed them techniques for being successful. During the meal, they discussed the transferable skill that each person brought to the table; this further cemented the capability of each person to be confident in their abilities. It dramatically changed their attitudes and thinking about taking on the job they didn't really want to do and helped them to become comfortable with doing it very quickly.

Pilot projects, on the other hand, are proof of concept initiatives that run over a longer duration than quick wins, but are no more than 50% of the implementation effort. They are utilized to help refine the new products and services and to help people become confident in using them. Pilot projects can start at any time during the implementation and can be considered a *soft implementation*.

Pilot Project Example

An example of a pilot project is the *soft opening* before the grand opening of a new store or restaurant. In the case of a soft opening, the restaurant actually opens up to a month in advance of the grand opening. During that time, staff are hired, trained, and the overall services and menu items are tested and refined.

How Do Quick Wins and Pilot Projects Impact the Ability to Change?

Again, quick wins are designed specifically to cause a shift in culture, thinking, and the emotional response to impending change; while pilot projects are designed to test out the process itself. To that end, quick wins create a shift by demonstrating what success looks and feels like to those impacted by change. It makes them feel as though they can be successful using the new products or processes, whereas pilot projects simply apply the new products and processes in order to identify areas for improvement prior to the full go-live.

Both are great tools when they are well-designed and leverage appropriate cases and opportunities. However, either one can have the exact opposite effect if not done well.

Quick wins have the ability to turn around negative attitudes and resistance to change in a very short period of time. However, in planning change efforts, do *not* plan only one or two quick wins and expect the result to be long-lasting. To leverage quick wins as a tool, the change strategy and plan must contain several quick wins that target key milestones and types of resistance. In other words, you have to choose the appropriate activity for the appropriate audience; and schedule that activity to occur at the right time.

Pilot projects have the ability to build confidence over a longer period of time as people learn and develop a level of comfort from using the products or processes.

Using quick wins with pilot projects is important because people with negative attitudes can go into a pilot project and sabotage it. Because of the nature of quick wins, they will be a valuable asset in shifting those negative attitudes before the pilot project kicks off. In fact, the best way to add value and take advantage of the pilot is to ensure that the participants and stakeholders go in with a *can-do* attitude as opposed to the *it'll-never-work* attitude. When glitches in the pilot project occur (and make no mistake, glitches will occur), the participants with the can-do attitude will actively work to refine the process or the product through problem solving, instead of taking it as proof that they were right all along and then convincing everyone around them to throw in the towel.

The bottom line is this: you cannot employ a pilot project without quick wins (at least not without significant risk), but you can employ quick wins without leveraging a pilot project.

PROVIDE ONGOING SUPPORT

One of the most critical aspects of successful change efforts is that people must feel as though they are not alone when they are going through it. This isn't just about how much effort it takes to turn a ship; it's about how people feel about change. The analogy about turning the ship is really what happens when you magnify the effort that it takes to help a single person cope with and navigate change by the total number of people in the organization. In this case, a better analogy is the *imagine you've suddenly gone blind* scenario discussed earlier in this chapter.

Again, imagine you were going for your typical walk around the neighborhood and suddenly could not see anything. What would you do? Most likely, you would be startled, fumble around for a couple of steps, stop dead in your tracks, and be paralyzed by fear. Would you remember the way without being able to see it? Could you give directions to another person to help? Or, would you be so overcome with fear that you would feel totally and completely lost despite having walked this very route a thousand times?

This is exactly the reason that managing change is about providing support to the people impacted by it. In Chapter 1, we talked about the barriers to change. Support—ongoing, emotional support—is designed to help them overcome these barriers.

PROVIDE POSITIVE REINFORCEMENT AND GOVERNANCE

While support before and during change is crucial to the success of the initiative, it is not the only factor that must be in place to ensure that adoption of the change occurs and lasts. Support as a mechanism for enabling adoption is one part of an equation. The other parts of that equation include positive reinforcement and governance.

Support is not the same thing as positive reinforcement. Where support is the feeling that a person has of being a part of a team effort and that others on that team *have their back* so to speak, positive reinforcement is when desirable benefits are offered in exchange for demonstrating the specific behaviors or performing work in the manner identified as the *chosen method*. Positive reinforcement has a goal of increasing how often desirable behaviors are demonstrated or the work is performed according to the chosen method.

> ### Example of Positive Reinforcement
>
> An example of positive reinforcement is a tangible reward. A bank has a rewards and recognition program that specifically provides small gifts to any bank employee who obtains a perfect score from their secret shopper. When the person obtains four or more perfect scores, the bank provides gift certificates for dinner and movies.

In the age of online transactions, more and more companies are leveraging gamification to provide this reinforcement. What is gamification? In a nutshell, it is leveraging the statistics from clusters of people that are similar to you (as the user), to turn the use of tools and processes into a competition between peers.

That being said, successful change initiatives also require governance. Governance is the framework whereby adherence to processes is managed and measured, but it also provides the guidelines for consistency in delivery and application. It enables consistency in key decision making and removes subjectivity and personal bias to the degree possible as determined by the due diligence and rigor around the control gates.

DEMONSTRATE THE EFFECTIVENESS OF THE CHANGE EFFORTS (BEFORE, DURING, AND AFTER)

One thing that far too many transformation initiatives fail to do is to demonstrate the effectiveness of the changes consistently across all stages of the effort. Successful transformation programs showcase and advertise their progress regularly.

While many teams are really good at showcasing progress during the initiative, they forget to talk about it *before* it happens and *after* the project has wrapped up. However, if the progress of adoption is dropped as quickly as the project wraps up, many people will go back to business as usual before the changes have been fully cemented—and this is dangerous because the organization is at risk for a long, drawn-out, and incomplete adoption. People need to know how great the changes are, and they need to hear how others they work with are being successful when it comes to applying the new processes and leveraging the new tools.

REFERENCES

1. Thorndike, E. L. (1911). Provisional Laws of Acquired Behavior or Learning; Animal Intelligence (New York: The McMillan Company); Wolpe, J. (1968). *Psychotheraphy by Reciprocal Inhibition.* Conditional Reflex 3 (4).
2. Kübler-Ross, Elisabeth, *On Death and Dying;* Routledge, 1969.

5

Understanding the Business

The organization is comprised of two equally important parts: the ecosystem and the climate (shown in Figure 5.1).

UNDERSTANDING THE BUSINESS ECOSYSTEM

Before any change initiatives can be successfully undertaken, especially in the case of short-term change consultants, it is extremely important to understand some fundamentals about the overall business and how it

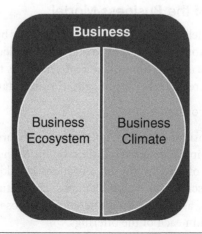

Figure 5.1 Business composition

operates. The basic elements that must be understood are the overall business model, the key relationships of the business (internal and external), and the organizational structure.

It is the combination of these basic elements, coupled with the knowledge of the business climate and the details of the specific changes, that will enable the successful planning and execution of the transformation. This knowledge leads to greater success because it ensures that the planning can accurately predict all of the areas that will be impacted and how they will be impacted by the transformation. It further ensures that all stakeholders are identified early on so that appropriate communications can be distributed at precisely determined times throughout the transformation. All of this ensures the highest possible levels of engagement in the overall transformation of the business.

BUSINESS MODEL

A business model, as illustrated in Figure 5.2, is a logical construct that summarizes the *core interrelated architectural, co-operational, and financial arrangements designed and developed by an organization presently and in the future, as well as all core products and/or services the organization offers, or will offer, based on these arrangements that are needed to achieve its strategic goals and objectives.*[1] In other words, business models describe how organizations capture, create, and deliver value across economic, social, cultural, and other applicable contexts. Constructing business models is part of strategic planning.

Dimensions of the Business Model

The business model describes what the business does by seeking to answer these basic questions: what products and services does the organization sell; how does the organization make money; who is their target market; and how does it create value. Specifically, the important dimensions are:

- Value Proposition

The value proposition dimension (shown in Figure 5.3) asserts that business models include an overall description of the products/services offered or to be offered by the organization. In addition, the business models must describe the value elements, the nature of the targeted market segments, as well as the preferences of those segments that are incorporated within each of the offerings.

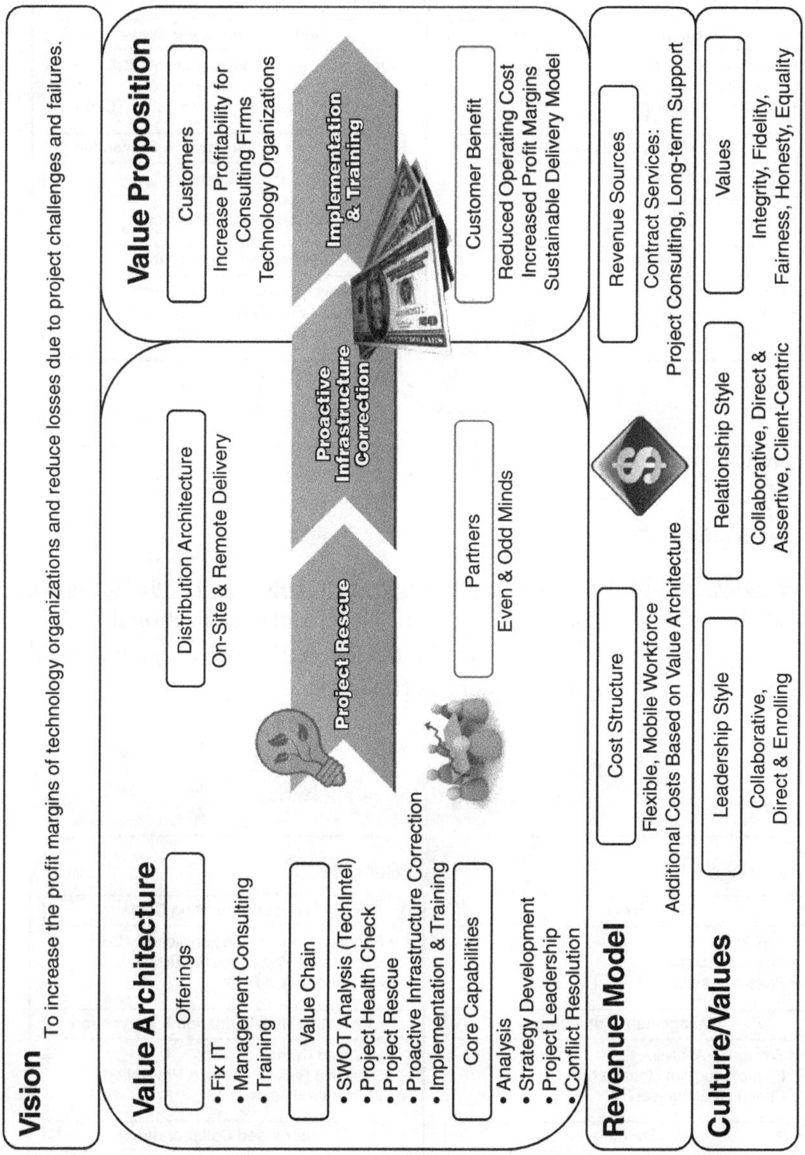

Vision To increase the profit margins of technology organizations and reduce losses due to project challenges and failures.

Value Architecture

Offerings
- Fix IT
- Management Consulting
- Training

Distribution Architecture
On-Site & Remote Delivery

Value Proposition

Customers
Increase Profitability for
Consulting Firms
Technology Organizations

Value Chain
- SWOT Analysis (TechIntel)
- Project Health Check
- Project Rescue
- Proactive Infrastructure Correction
- Implementation & Training

Project Rescue

Proactive Infrastructure Correction

Implementation & Training

Partners
Even & Odd Minds

Customer Benefit
Reduced Operating Cost
Increased Profit Margins
Sustainable Delivery Model

Core Capabilities
- Analysis
- Strategy Development
- Project Leadership
- Conflict Resolution

Revenue Model

Cost Structure
Flexible, Mobile Workforce
Additional Costs Based on Value Architecture

Revenue Sources
Contract Services:
Project Consulting, Long-term Support

Culture/Values

Leadership Style
Collaborative,
Direct & Enrolling

Relationship Style
Collaborative, Direct &
Assertive, Client-Centric

Values
Integrity, Fidelity,
Fairness, Honesty, Equality

Figure 5.2 Business model

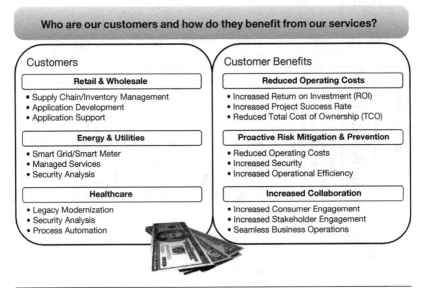

Figure 5.3   Value proposition

- Value Architecture

The value architecture dimension (shown in Figure 5.4), on the other hand, illustrates the concept of a holistic structural organizational design. This includes the appropriate technological architecture, organizational infrastructure, and applicable configurations.

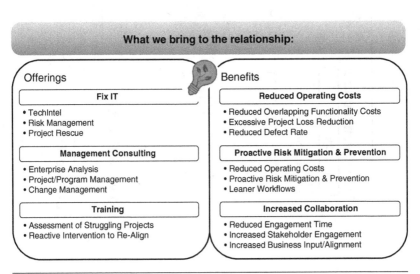

Figure 5.4   Value architecture

- Value Network

The value network dimension depicts the cross-company or inter-organization perspective toward the concept. In other words, it describes how the organization interacts with complementary partner organizations to deliver the values described within its business model.

- Revenue Model

The revenue model dimension (shown in Figure 5.5) illustrates information related to cost models, pricing methods, and revenue structure of the organization.

The business model is important within the context of change management because it provides specific information to the change agent about the organization, its priorities, and value drivers. This information can be leveraged to understand and promote change within that organization. Additionally, these models are leveraged internally by managers in order to explore possibilities for future organizational development.

Design logic, on the other hand, perceives this model as a by-product of either creating new organizational structures or changing those existing structures in order to pursue new opportunities. It goes without saying that these new opportunities generate transformation efforts in order to bring them to fruition. In reality, the influence of business models on change management is two-fold: it informs the change agent of the organization's value drivers and it recommends the changes being made.

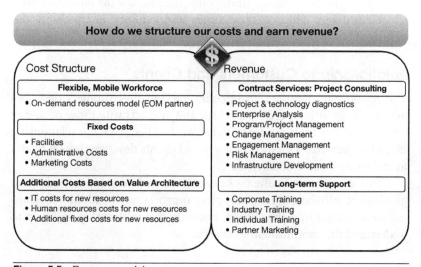

Figure 5.5   Revenue model

Gerry George and Adam Bock[2] conducted a large-scale review of existing literature, and further surveyed managers in order to understand how those managers perceived the elements of business models. They were able to demonstrate that there is some degree of design logic behind the way in which entrepreneurs and managers perceive and articulate their business models.

In extensions of the design logic, George and Bock[3] leveraged case studies coupled with IBM survey data on business models within large companies, as a means of describing how chief executive officers (CEOs) and entrepreneurs created stories to transition the organization from one opportunity to another. They also illustrate that when that story is either incoherent or the components of the story are misaligned, these organizations tend to fail. In the end, George and Bock recommend ways in which the organization can articulate powerful narratives for change.

KEY RELATIONSHIPS

Some aspects of working within a company are all about the big picture. Organizational change and transformation is one of those aspects. That means that in order to plan for the transformation, change agents really do need to understand the business as a whole. There may be risks and issues hidden in the minutia of these details, outside of what is considered to be scope, that could have detrimental impacts on the project and its outcomes—like adoption rates. Therefore, it is imperative that people looking to generate a change strategy understand how the business relates to its customers and clients, partners and vendors, competition and industry.

Relationship to Customers and Clients

The relationship between an organization and its customers describes how an organization perceives, values, and interacts with those people or entities that buy their products and services. Effectively, the relationship that a business has with its customers and clients describes how it relates to and treats those that it serves.

This relationship is often a key indicator of leadership attitudes that give insight into the culture of any given organization. In addition to gaining some insight into the culture, knowing this relationship will help to understand the commitment to change.

> ### Customer Relations
>
> Consider a small organization that is planning a major initiative to grow its customer base by adding a new service line. The relationship that the organization has with its existing customers is contentious because the customer service for existing products and services is poor.
>
> Realistically, how well do you think the leadership of this company is committed to making the changes to support customers when they feel as though some customers are *always complaining about everything*?

In the above scenario, it is easy to see that the organization has a long way to go to make the changes because the customers probably don't really want those new services; they only want better customer service. Leadership is not as committed to adding new services for customers who don't appreciate them and what they do. So how successful can this change be?

Understanding how people change is only half the battle. Understanding *how* and *how well* change will be adopted over what time period is the other half.

Relationship to Partners and Vendors

The relationship that an organization has with its partners and vendors describes how the business finds the resources needed to provide its products and services to its customer base. The relationship between a company and its partners and vendors is not always positive. In fact, sometimes, just as it is with customers, that relationship can be full of mistrust and abuses.

There is a power balance or imbalance between the organization and its vendors and partners. How it navigates that power and maintains it is crucial toward understanding how that same organization views its own people. There are organizations that think that they are in competition with their own vendors, and even their own employees.

> ### Internal Competition
>
> Several years ago, there was a large consulting firm with a manager that pushed his contract employees to produce. These contract employees were from several partner companies that held agreements to support the success of the consulting firm.
>
> Unfortunately, when these contract employees produced, the manager critiqued the work and ripped it apart. Now, had he left it there and simply asked the employees to redo the work, it would have been one thing—but this manager insisted that the employee redo the work, while the firm's employees did it as well. He insisted that the client should choose between the deliverables in hopes that this would prove the firm's superiority.

In the previous scenario, when faced with the selection, the client was confused at first, and then irate. They were not paying for this kind of thing. This change project required a lot of change management to support the transformation. Mostly, time was spent managing the relationship between the consulting firm and its own partner rather than in managing the changes being made.

It is important to understand how the organization views its own partners and vendors, in the same way that it is important to understand how the organization regards its customers, because it helps the change agent understand the power leveraged within that same organization. It also helps to further cement how leadership responds and interacts with others as an indicator of how well the changes will go.

This relationship is a really good measure of how well an organization feels about its own employees, since partners and vendors are often treated as an extension of their own internal resources. Negative attitudes could mean a disinterested and disengaged employee population, and this will cause the transformation to sputter and drag on, if not stall outright.

Relationship to Competition

Every organization in the world has some form of competition because there is always another organization out there working to provide a similar set of products and services to the same customer base. By and large, a lot of organizations have traditionally viewed the competition as *the enemy*, and this has been reflected in their culture—their interactions with clients, partners, and vendors.

The world is changing and this is no longer the norm. The competition is no longer the enemy and those that interact with them are no longer seen as traitors.

Shopping

Many people can relate to going to find something very specific and not finding it at the store they were sure would carry it. Years ago, it was considered a fireable offense to advise a customer to go to another store to find it, even when this store did not even sell it. Thankfully, many stores realized that they won more customer loyalty by telling them where to find the products than by keeping that information to themselves.

First and foremost, this is a sketchy view of a fellow organization working to provide useful services and products to those who need them, but it is also a short-sighted view of your (potential) customers; insinuating that they are a commodity and to be taken for granted, rather than sentient beings who are capable of thought. Consider how this organization would respect these customers when it comes to changes that impact them. They are probably not going to listen to the feedback provided.

The relationship that an organization has with its competition really describes two things: how they see themselves and what the leadership values in performance.

Now, what if they not only see another organization as the competition, but also another internal department of the same organization? How well can change be expected to progress or be implemented when they feel as though other departments are *the enemy*? This is clearly not good for implementing change because people are more likely to be in conflict with one another than they are willing to work together.

Relationship to Industry

The relationship to industry describes how the organization sees itself in terms of the industry it serves. That is to say, the relationship to industry describes how the organization sees itself compared to the industry norm. The industry norm is the set of best practices being utilized to establish how individual organizations within the industry deliver products and services to its client base, and the standard measures being applied to assess organizations against that standard.

This view is important because organizations have to have a realistic view of how well they are applying the best practices and how well they deliver against a common benchmark. When organizations do not have a realistic view of themselves against this standard, it can be very difficult to convince them that change is even necessary.

Big words do not always equate to big action, and the people impacted most by change know this. Where everyone else might see a strong company, the people on the inside know its strengths and weaknesses and know if it is generating a realistic picture or not. When an organization does not generate a realistic picture of itself and the vision does not align to reality, people in the organization will have a harder time accepting change.

ORGANIZATIONAL STRUCTURE

Organizations typically organize themselves into the best possible structure for meeting their objectives, and often this is replicated from similar entities within a similar industry. However, structure can also say a lot about those running the organization, and in turn, this reveals a lot about how change is internally accomplished.

An organization's structure is an excellent indicator of the true values of the organization as a whole. There are three ways to organize any group or business entity that are important for this conversation about change: (1) to meet business strategic objectives; (2) to meet the ego of the leadership; and (3) to react to the world around it.

When an organization is set up to meet its own objectives, the business, and the people that work there, along with those who interact with it, everyone is focused on common goals. They adapt more readily to changes and have an increased sense of trust in the organization because their sentiment is that the organization is trying to do the best job possible for as many people as possible.

However, when an organization is set up to meet the ego of leadership, those who work with and interact with the organization are more likely to be in it for themselves. There will be higher amounts of interpersonal conflict. People feel taken advantage of and are less likely to adapt to changes because of the low levels of trust.

Finally, when an organization is structured as a response to the world around it, it is shape-shifting, chaotic, and routinely changes its structure. There is no real structure. People within are likely to be less committed to change because they know it will not be long before something new comes along to replace it. These organizations tend to have a high attrition rate.

UNDERSTANDING THE BUSINESS CLIMATE

The business climate can effectively be described as the environment in which the business employees operate on a daily basis. While one could argue that is the definition of culture, the truth is, that culture is merely a part of the climate. This is because there are other factors that contribute to the creation and reinforcement of culture within the business ecosystem. These factors include vision and mission; whether an organization is performance-based or not; the culture fostered by leadership; whether or not an organization is profit-based; customer loyalty; and brand strength.

Each of these factors will contribute to how well the people impacted by the proposed changes will react, recover, and adopt those changes, whereas culture will describe how they will interact with one another and reinforce reactionary behaviors.

Climate versus Culture

Within a moderately-sized financial services company, the president cultivated a climate of fear by bullying employees for not producing, making small errors, and not working excessive amounts of overtime. He bullied them by letting them know their job was on the line and, consequently, they lived in fear.

However, the culture that resulted from this climate was one where people had almost no loyalty and did not work well as a team. There was constant in-fighting. People were manipulative and willing to throw someone under the bus (as it were) when mistakes were made, and the president demanded answers. Ultimately, the climate of fear created and fostered a culture of *dog-eat-dog* and *every man for himself.*

BUSINESS VISION/MISSION

The vision is a clear picture of how the business will look, feel, and operate once the transformation has been completed. The mission, on the other hand, is the statement for how it will achieve the vision.

Aside from the fact that many businesses often blur the lines and confuse the two, many projects do not leverage these two crucial elements. Whenever change management is required on a project, both vision and mission statements are strongly recommended, however, a vision statement is absolutely necessary.

Vision Statement

A VISION STATEMENT IS ABSOLUTELY NECESSARY WHEN CHANGE MANAGEMENT IS REQUIRED.

Here's why: The fundamental element of change management is communication of a shared objective in order to rally the troops and engage the stakeholders and impacted groups. This shared objective is best communicated and embraced when it takes the form of a vision statement. Without it, change management is more difficult and stakeholder engagement over the long term is more difficult to maintain.

A mission statement, while not absolutely necessary, is still recommended because it directs the actions of the stakeholders toward the project by letting them know how they can help even if they are never directly contacted by the project team. It creates a sense of camaraderie and moral support for the project which actually fosters a social norm that encourages people to fall in line and adopt the changes. In other words, it's psychological.

PERFORMANCE-BASED ORGANIZATIONS

A performance-based organization is one that evaluates and judges the value of individuals strictly on their performance. These organizations do not consider or value other contributions because they do not understand how to value them or why they are important to the health of the overall team. People, in effect, are reduced to a rubric.

Example

Companies such as real estate firms, which routinely cut the bottom few performers as a means of motivating staff, are an example of this kind of organization. The financial services company in *Climate versus Culture* is also an excellent example of a performance-based organization.

Ironically, quality has a tendency to suffer as a result in a performance-based organization. This is due in part to the fact that when people do not feel important, they stop caring about detail and quality, and focus more on meeting the rubrics which tend to be concentrated on how many products a person can produce and how much money they can save while doing it.

The truth is, that while performance-based organizations may be perceived as looking good on paper (i.e., the bottom line), when you dig a little bit deeper, they have a fatal flaw. The people within these organizations are less likely to be loyal and are certainly more resistant to change. This is specifically because while these people can be good at producing and flying under the radar so that they are not terminated, they are not exactly happy either.

The level of job satisfaction among people within these organizations tends be lower overall and they tend to live in fear of losing their jobs. In fact, the higher the push for performance, the higher the level of fear. Why? Because each of these people knows that the moment their performance slips or someone comes along that performs better than they do, they are at risk of losing their job or at the very least, their hard-earned seniority.

Level of Job Satisfaction

The level of job satisfaction is the degree to which a person feels content with the job they perform day-in and day-out. This can be influenced by a variety of factors including the type of work that they do; the person they work for; the vision of the company; the reputation of the organization; the people whom they work with; and how important they feel to the work being done.

How Does Job Satisfaction Influence Culture?

Job satisfaction plays a pivotal role in the formation of culture because culture is a reflection of the values of the organization and the attitudes held by the people who work there. This will shift as leadership changes or new policies and procedures are put into play. The truth is that while people with low levels of job satisfaction tend to be the most afraid of change, they are also often secretly hopeful that things will change to increase their overall job satisfaction.

Bad Relationships

There is a reason that people stay in bad and abusive relationships. Ultimately, it is the hope that one day it will get better and that the situation will change to something more positive and equitable. Unfortunately, the data clearly shows that this rarely happens.

How Will Job Satisfaction Influence Change Reactions?

Just because people are secretly hopeful that things are going to change for the better, does not mean that they will support change or even embrace it. Fear of the unknown is often a far greater factor than the minute possibility of having a higher level of job satisfaction. People will still react negatively to change if it is not well-managed. Worse still, is that they will lose trust and faith every time changes are poorly implemented.

How Will Job Satisfaction Influence Recovery from Change Shock?

Change shock is the sudden feeling of shock that accompanies transformational activities, even if for a brief moment. People with a high degree of job satisfaction will recover much more quickly from change shock than those with a lower degree.

Imagine Jumping into the Driver's Seat

Imagine that you're the passenger in a car driving at cruising speed along a mountain road when suddenly the driver goes limp and passes out. You have to take over control of the car quickly or you'll crash! Wait, what do you do? Where are the controls? Where can you pull over?

In the above example, you start to go into shock because of the sudden change in roles. You took for granted that the driver was in control so you could just sit back and relax. You did not have to worry about anything—until the roles suddenly changed. It took a second to recover because you do, after all, know how to drive; you just were not prepared to be in that role.

How Will Job Satisfaction Influence Adoption?

Job satisfaction is a critical indicator of the rate of adoption that new changes will take to be fully implemented. Remember that scenario of the driver passing out? What if the passenger was suicidal and angry with the driver? It paints a whole different picture of how quickly and how likely they would be to jump into the driver's seat and take control of the car.

Adoption is really all about how well and how quickly people take on the new tools and processes and implement them into their daily routines. If they have low levels of job satisfaction, they are more likely to be checked out and not care about adopting new processes or tools. While they may not actively try to prevent the changes from being implemented, they won't exactly embrace them either.

How Can Transformation Initiatives Manage Change Without Job Satisfaction in a Performance-Based Organization?

The key to successfully implementing changes and transforming a performance-based organization with low levels of job satisfaction is to leverage as many change techniques as practical and to transform the relationship between leadership and the workforce. This relationship is crucial because if the change agent ignores it and simply implements the changes, people may adopt them slowly, but will eventually revert back to bad habits and in the long run, their job satisfaction is going to decrease further.

However, in order to transform this relationship, it is important to guide leadership into new management methods and to show them how to value people for their full contributions and not just their productivity. In other words, the change agent must teach leadership to trust and believe in the people who work for them. Leaders' implicit followership theories (*LIFTs*) are an extremely effective way to accomplish this.

LIFTs

According to Thomas Sy, LIFTs, "are cognitive categories that reflect the conceptions that leaders have about the traits and behaviors of followers."[4] In other words, LIFTs are the beliefs that leaders have about the characteristics and behaviors of their followers. These beliefs influence how the leaders treat these followers and then, in turn, influence how the followers respond and perform. Ultimately, it becomes a cycle of self-fulfilling ideals.

Because LIFTs are fairly stable ideas of the followers, and performance expectations are most likely to be heavily influenced by the follower characteristics within a specific situation, LIFTs can help leaders to create and hold specific performance expectations across multiple situations and followers.

Different Situations, Different Performances

Have you ever known someone who rants about another person and how they're going to give them a piece of their mind the next time they see them? And then when they actually do see them, they are meek and quiet.

Well, in different situations, the person doing the ranting may feel more powerful. Maybe their friends are around when they are venting, and maybe their grandmother is around the next time they see the person they are venting about and don't want to upset her so they keep quiet.

In the above scenario, it is clear how the situation governs the behavior of the people in it. Based on what has been discussed so far about LIFTs, it is easy to see how the leader can gain an understanding of how a person behaves in a particular situation, and come to gain performance expectations for that person in specific situations.

Realistically, positive LIFTs are the perceptions that leaders hold for their followers, and it's important to note that these perceptions are distinct from expectations. When those positive LIFTs are *activated*, expectations related to performance are also more likely to become activated at the same time. What that means is that positive LIFTs influence leaders' expectations of their group of followers for the better.

CULTURE

As will be discussed in Chapter 6, culture is the social and interpersonal environment of the organization. It is determined by the management and reinforcement of key values and behaviors among the teams within that organization. However, culture is heavily influenced by the aspirations of leadership.

Aspirations are those ideals and goals that have been set by the leadership for where they would like the organization to end up. That being said, aspirations can actually have a demotivating effect on employees and create a toxic culture when they do not align to, and are not reinforced by, the values espoused by those same leaders. This is because values are exhibited through behaviors.

For example: A leader might aspire to have an open door policy and to have a team of direct reports that come to them with their concerns. But that same leader values people who are go-getters and work completely independently. The behaviors that are most likely to be exhibited in his team are going to be the exact opposite of his aspirations. They will not approach him, and there will likely be a breakdown in communication.

Toxic Culture

A wholesale company had a bad reputation for the infighting and bullying which occurred on a daily basis among employees. In one particular incident, the regional manager posted a derogatory notice about one of his direct reports in the monthly newsletter and distributed it for all to see.

The regional management often laughed off a bullying accusation and frequently promoted others who bullied particular employees. It was a horrible place to work, and everyone knew it. This particular company had a very high attrition rate.

Desirable Culture

An insurance company had a fairly good reputation for how it felt to work there and for how employees felt they were treated. Overall, employees were permitted to choose to either work from home or in one of the company's offices. People promoted collaboration and support of other team members and generally behaved like they enjoyed working together.

The company not only espoused these values, but also promoted people who displayed them. By employee standards, many of them enjoyed the culture so much that they had a high tenure rate and low attrition rate.

It can be relatively easy to spot both a toxic and a great culture by the types of behaviors that are displayed at the differing levels within the organization. However, not all organizations in the middle of the spectrum are as readily identifiable, which makes them more difficult to change and transform.

Level of Intimacy

Level of intimacy can best be described as steps in the process that people go through in getting to know one another. As shown in Figure 5.6, these steps include: safe communication; others' opinions and beliefs; my opinions and beliefs; my feelings and experiences; and my needs, emotions, and desires.

Level One: Safe Communication

Level one is the lowest level of communication, and it involves the simple interchange of either essential or trivial information. It is impersonal with little risk of rejection or exposure of vulnerabilities and, therefore, it is considered to be *safe*. Topics that are a part of so-called *safe communication* often include discussions on the weather and traffic.

Level Two: Others' Opinions and Beliefs

Level two is when people begin to share more about themselves and is characterized by the sharing of thoughts and feelings, beliefs, and opinions. Interestingly, people are actually sharing the opinions of others they agree with instead of their own. It is believed that people are looking for synergy with the other person by doing this. At this level, topics might include reactions to a newscast or television show.

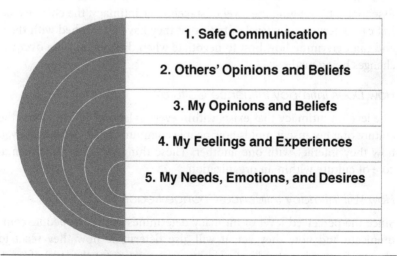

Figure 5.6 Levels of intimacy

Level Three: My Opinions and Beliefs

It is at level three that people begin to take more risk by revealing their own personal beliefs and opinions. It is risky because they are opening themselves up to criticism and potential rejection. It is at levels two and three that people are most likely to hide or alter their opinion if they feel threatened or the need to avoid conflict.

Level Four: My Feelings and Experiences

By level four, people are more open to sharing their own feelings and experiences. For most people, this is a very vulnerable position to be in because they discuss their successes and failures, as well as their dreams and goals.

Level Five: My Needs, Emotions, and Desires

Level five is the highest level of intimacy and it is the level where people are known for who they really are. This is because they share what is most important to them—their needs, feelings, and desires—and this level requires the greatest amount of trust. When people feel like there is still a chance of rejection, they cannot trust the other person with their deepest, darkest secrets.

Why Is Intimacy Important?

True intimacy takes time. As a change analyst, it is important to understand that people will behave and communicate based on their level of intimacy with you personally, as well as with others on the team. Essentially, by gauging the level of stakeholder intimacy, the change analyst can determine the level of trust that they have established with them and can determine how best to negotiate when differences arise over the changes being made.

How Does Intimacy Influence Culture?

The level of intimacy that exists within every relationship influences the culture of it because it molds how people communicate, behave, and even how they interact with one another. These things are the fundamental core of culture within an organization.

How Will Intimacy Influence Change Reactions?

Since the perceived level of intimacy determines both how people communicate and how they act, it will also determine how they react to change. Remember that level of intimacy is a gauge for the level of trust.

People are more likely to react negatively when there is a low level of trust in the relationship. One key thing to keep in mind is that intimacy is measured by the person with the lower level of trust.

Level of Trust

A Definition of Trust

"The degree of confidence born of the character and competence of a person or an organization." —Stephen R. Covey

According to Covey, there are 5 Waves of Trust: self, relationship, organizational, market, and societal (shown in Figure 5.7). Each of these waves is important in developing and maintaining relationships with others, therefore, it is important to understand these different waves.

Self

Self-trust is all about credibility. It begs the question, "Am I someone that everyone, including myself, can trust?" In other words, "Do I have

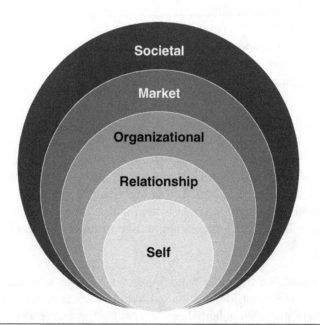

Figure 5.7   Covey's 5 Waves of Trust

credibility." To answer this question, Covey calls out what he identifies as the four cores of credibility. These are:

- Integrity
- Intent
- Capabilities
- Results

Relationship

Relationship trust, on the other hand, is all about behavior. It is about the 13 specific behaviors that people demonstrate on a consistent basis in order to garner trust from others in relationships. According to Covey, these are:

- Straight talk
- Demonstrating respect
- Creating transparency
- Righting past wrongs
- Demonstrating loyalty
- Delivering results
- Improving or demonstrating improvement
- Confronting reality
- Clarifying expectations
- Practicing and demonstrating accountability
- Listening before speaking
- Keeping commitments
- Extending trust

These will be discussed in greater detail later in this chapter.

Organizational

Organizational trust concerns trust with *internal* stakeholders—the people within the organization. It is the perception of those within the organization that it will do right by them, or at least the extent to which those people trust the organization to support them in times of need.

Market

Market trust, on the other hand, describes the trust with those stakeholders that are outside of the organization, such as vendors, partners, customers, and even potential customers. It is the degree to which these people outside of the organization trust it to do the right thing when faced with any ethical or operational dilemmas.

Societal

Societal trust is the trust that is earned when the organization contributes to the world at large. In effect, it is a form of social currency that can be leveraged when the organization requires support.

A Note About Trust

There are many events that can break the trust between any two parties, but only a few things that will bring it back. However, the ability to restore trust depends largely upon three things: the nature of the loss, the importance of the relationship, and the willingness of both parties to restore it.

That being said, when others have lost your trust, it is important to do the following in order to work at restoring that bond:

- Be quick to forgive them and think about the situation from their perspective
- Do not be quick to judge the situation until hearing the facts from that perspective
- Do not overthink it and start considering a hidden agenda or conspiracy

If, however, you are the person who has lost the trust of others, take the following steps to repair the damage:

- Work hard to strengthen the core elements of trust (integrity, intent, capabilities, and results)
- Behave in ways that inspire trust from other people
- Be willing to work without a hidden agenda and be transparent

If trust truly is the confidence that arises out of character and competence, then trust is a mandatory requirement in the process of transformation. It is as fundamental to change as the alphabet is to reading. Change is all about moving from one state in the here and now, to something heretofore unknown. Successful change is about ensuring that people adopt quickly and shift when required.

Coming About

In sailing, when the boat is going to come about, all passengers and crew are required to shift their weight at exactly the right moment in order to ensure that it does not capsize.

Effective coordination and leadership are critical to successful change. In order to have and maintain that coordination and leadership, and to have people listen when those leaders ask people to make the changes (i.e., the shift), the people being asked must trust the leader and the organization.

How Does Trust Influence Culture?

The truth is that culture is synonymous with trust. Since culture is the set of shared beliefs, practices, and behaviors that are considered socially acceptable, and the level of organizational and personal trust heavily influences how people act and respond, it stands to reason that culture is highly indicative of the level of trust and vice versa.

How Will Trust Influence Change Reactions?

The level of trust within an organization actually sets the tone and the precedence for how people will react and respond to change. Again, if the definition of culture is to be accepted, culture that is heavily influenced by change has standard sets of responses and behaviors that people will utilize when changes are announced.

That means, where people have low levels of trust in the organization and its leaders, they are likely to have behaviors that are negative or passive-aggressive. When it comes time to transform parts of that organization, those people who are directly impacted by the changes will revert to those behaviors in response.

How Will Trust Influence Recovery from Change Shock?

Change shock is the initial reaction that people have between the time that changes have been announced and the time they have had to mentally process the information. This time frame varies for every individual, but in reality, *how* change is announced will have an impact on this, but so will the level of trust that people have.

As with the reactions to change, the recovery from change shock will be faster when there is a higher level of trust. Conversely, the recovery process will take longer when people have lower levels of trust either in the organization or their leadership.

How Will Trust Influence Adoption?

In the same way that people will react to change and need to recover from change shock, they will adopt the changes based purely on how much they trust their leadership and the organization. Lower levels of trust equate to a longer time to adopt the changes.

The real question here is: can transformation initiatives be managed effectively without trust in performance-based organizations—and how? The short answer is: yes—by employing the various approaches and techniques outlined in this book, change can happen in even the most difficult organizations.

Behaviors that Build Trust

A few pages ago, I listed Covey's 13 behaviors that build and foster trust within relationships. In more detail, these are:

- Straight talk—clearly articulating your needs and wants to the other party in a truthful and tactful manner, while leaving nothing to be assumed
- Demonstrating respect—showing the other person respect for them, their integrity, and their dignity
- Creating transparency—creating opportunities to discuss things openly and clearly so that everyone is on the same page and there are no surprises late in the game
- Righting wrongs—taking time to recognize errors that have been made and working to make amends
- Showing loyalty—honoring your relationships as separate and unique by not gossiping or sharing information out of turn, and standing up beside those in the relationship
- Delivering results—turning requests and suggestions into tangible outcomes that can be leveraged in the transformation effort
- Improving behaviors and performance—working to improve on the performance by learning from mistakes and asking for constructive feedback, then acting on that feedback
- Confronting reality—tackling issues head on and diving straight into discussing them
- Clarify expectations—reaffirming what both parties in the communication expect to see in terms of outcomes, next steps, and so forth
- Practicing accountability—demonstrate a willingness to be accountable for the actions taken, decisions made, and words spoken
- Listening before speaking—taking time to listen to the other party and clarifying what they have said before formulating a response

- Keeping promises—ensuring that expectations are managed and commitments are kept
- Extending trust—offering to trust before evidence of any judgments or past actions enter into the picture

PROFIT-BASED ORGANIZATION

Any organization that solely determines success based on the level of profit and is responsible only to its shareholders is profit-based. As with the performance-based organization, this one is also run and controlled using a completely top-down business model.

It is extremely important for the change agent to know the type of organization they are working with because there will be crucial differences in how people respond to and adopt change. These crucial differences will come directly from the attitudes that they personally have for the organization and the work that they do.

According to Phillip C. Thomas, there are three attitudes that segment market trust—*promoter, passive,* and *detractor.*[5] These attitudes are also not only characteristic of an organization's consumer, but they are equally prevalent among the people working within its walls. When it comes to change, they become more apparent depending on the type of organization.

The person with the promoter attitude actively promotes the organization's products and services and will go out of their way to do business with them. When it comes to change, this person is also known as the champion.

The person with the passive attitude, however, only uses the products and services out of convenience, and would be quite fine if they were replaced by another organization. When it comes to change, this personality makes up the largest contingent of the people impacted by the transformation in that they will really only adopt the changes because it is more convenient to do so.

The detractor, on the other hand, is the person who actively advises against doing business with a particular organization. When it comes to change, this is the person that the change agent should focus on converting to a promoter. The trouble is, that within performance-based organizations, there will be far more people with this negative attitude, and change becomes exhaustive and overwhelming.

The key to these attitudes is that they will appear in varying numbers and degrees within any given organization. However, the numbers and degree of the attitude is determined almost entirely by the type of organization in which these people work.

That means that while change agents will find varying numbers of promoters, passives, and detractors within any organization, they are more likely to find larger numbers of passive people within those that are profit-based and more detractors within those that are performance-based.

Understandably, the return to shareholders is almost three times greater for organizations with high internal trust than it is for those with low trust.[6]

How Can Transformation Initiatives Manage Change in Profit-Based Organizations?

It is important then to understand how to manage each type of participant based on the type of organization, and it is equally important to understand how that organization manages its people, and therefore, how to implement change. Change can be either top-down or bottom-up and as the change agent, it is crucial to know the difference, along with where and when to utilize the tactics to influence change.

BRAND STRENGTH

Brand strength can be defined in three different ways: current competitive performance; relevant beliefs and attitudes of the consumer; and estimating future performance and profit streams. For the purposes of this discussion, it is the *beliefs and attitudes of the consumer* that are important.

When it comes to change, it is important to understand that the brand strength of the company plays a role in the level of trust that exists between not only external consumers, but internal resources as well. The more favorable the beliefs and attitudes of internal resources, the more likely they are to believe in and trust the organization to do what is right. This ultimately means that they trust the organization to lead them through change safely.

How Does Brand Strength Influence Culture?

Brand strength influences culture in a very direct way. It increases internal loyalty and helps to shape the perception of how to communicate and behave within the organization. Remember that culture is about the practices, shared beliefs, and behaviors that are considered normal within the organization. This can absolutely be based upon the perception of the brand strength of the organization.

How Will Brand Strength Influence Change Reactions?

Change reactions are purely the initial way that people behave when they hear that change is coming. Surprisingly, when it comes to change reactions, people in an organization with greater brand strength are more likely to accept that change is coming and even to begin reaching out to participate.

How Will Brand Strength Influence Adoption?

As with culture, brand strength influences change adoption in a very direct way. More people are likely to roll with changes when the company has greater brand strength. One could argue that this is because of the higher level of innate trust that people within the organization have for the people and for the processes involved; but it could also be that people are less likely to want to *rock the boat.*

CUSTOMER LOYALTY

Unlike brand strength, which is more of an indicator of the motive behind repeat business, customer loyalty is really a measure of how likely customers are to buy the products and services of a given organization over those of the competition. Customer loyalty is based on relationships with the organization and not the perceived value of the brand.

In terms of change and change adoption, customer loyalty is often in tandem with employee loyalty. It is the demonstration of mutual respect between the organization and its clients, and therefore will reflect how well the change will be accepted, both internally and externally.

How Does Customer Loyalty Influence Culture?

In a nutshell, customer loyalty helps to foster organizational pride. It is that pride in the organization that has the greatest influence on culture. When people have pride, they behave, respond, and react differently. Since culture is a shared set of behaviors, it follows that customer loyalty influences culture by increasing or decreasing the amount of pride that people have within the organization, and ultimately, how much they are willing to do for it.

How Will Customer Loyalty Influence Recovery from Change Shock?

As with trust and intimacy, customer loyalty influences the recovery from change shock by increasing or decreasing the time to recover. Again, where there is a high level of trust, intimacy, and customer loyalty, there is more likely to be a faster recovery period.

Moreover, people within the organization are increasingly likely to feel personally responsible and accountable for supporting customers through the changes and will therefore be quicker to adopt the changes themselves. In the end, everything about the business climate comes down to culture; and that can be reduced to the discussion on organizational trust, intimacy, and pride.

REFERENCES

1. Al-Debei, M. M., El-Haddadeh, R. and Avison, D. (2008). "Defining the business model in the new world of digital business." In *Proceedings of the Americas Conference on Information Systems (AMCIS)*.
2. George, G. and Bock, A. J. (2011). The business model in practice and its implications for entrepreneurship research. Entrepreneurship Theory and Practice.
3. George, G. and Bock, A. J. (2012). *Models of opportunity: How entrepreneurs design firms to achieve the unexpected*. Cambridge University Press.
4. Sy, Thomas. "What do you think of followers? Examining the content, structure, and consequences of implicit followership theories. Organizational Behavior and Human Decision Processes"; 2010.
5. http://www.slideshare.net/robertrtei/the-five-levels-of-trust-that-drive-success-or-failure.
6. Human Capital Index; Watson Wyatt Worldwide.

6

Understanding the Change

There is a direct relationship between the business ecosystem, the changes being implemented, and the tactics required to manage that change throughout the transformation. Where the business ecosystem provides the context for change, the specific types of changes will dictate the activities and tactics that are selected to ensure stakeholder engagement.

It is as important to understand the changes being implemented as much as it is to understand the ecosystem of the business itself. However, in order to understand the changes, there are a few attributes that must be considered—the business goals and drivers, the people impacted, the processes, and the specific technology being changed.

Where the business goals and drivers can help the change agent understand the purpose of the changes, it is the demographics of the people being impacted coupled with the processes and technology being changed that will determine the reaction, and thereby will drive the change management strategy and activities.

BUSINESS GOALS AND DRIVERS

Every change happens for a reason, even if that reason is not readily apparent, justifiable, or rational. It is important to understand the business goals and drivers behind the change, even if they can be summed up simply as some new executive trying to making their own mark on the business.

115

Knowing the business goals and drivers for change enables the change management team to create a positive and engaging message that encourages the impacted stakeholder groups to get on board and adopt the change. Lack of communication does little if anything to draw people on your side and get them excited about adopting change. In fact, it has the opposite effect, and more often makes people feel as if the executive expectation is that they will do as they are told—end of story.

Only, it's never the end of the story. Have you ever watched a project implode slowly because it took way more time to build and adopt new processes than it should have?

CHANGE SCORECARD

Where the balanced scorecard is utilized to establish the initial planning baselines, the change scorecard is the primary mechanism for measuring the performance of change throughout the duration of the process of transformation. The sole purpose of the change scorecard is to effectively answer four very basic questions:

1. What is changing, by how much, and over what time period?
2. How difficult is that change going to be?
3. How critical is it for the organization to make the changes?
4. How ready is the organization to make the necessary changes?

In answering these questions, the change scorecard effectively provides a very detailed look at the changes within the context of the business; its people, processes, and technology. This is illustrated in Table 6.1. The change scorecard is discussed in greater detail in Chapter 10.

PEOPLE

From a change perspective, it is critical that the project teams and businesses understand and define how changes will impact people at all levels of the organization. According to *self-determination theory* (SDT), shown in Figure 6.1, people have three basic needs—autonomy, relatedness, and competence—which complement Maslow's hierarchy of human needs (shown in Figure 6.2). Figure 6.3 shows how these theories interrelate as the top of that hierarchy is further expanded by the addition of the needs described by SDT. It is important to understand how these concepts are

Table 6.1   Change scoreboard (the Likert scale is used for scaling responses in research)

Factor	Description	Score	Likert	Ind. Weighted Score	Max Weighted Item Score	Final Weighted Change Score
				100%	100%	100%
Degree of Change	% of change that must occur	68%	3	50%	20%	10.00
Complexity	How difficult the changes will be	High	5	100 %	25%	25.00
Criticality	How important this change is to the business	Important	3	50%	30%	15.00
Business Readiness	How prepared the business is to tackle them	95	5	100 %	25%	25.00
						75.00

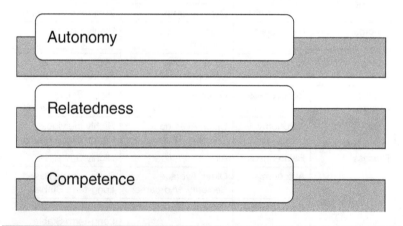

Figure 6.1   Self-determination theory

related so as not to misunderstand the importance of each. Anytime change impacts one or more of these needs (in varying degrees), there is a propensity for people to react negatively (see Table 6.2).

Further, Figure 6.4 illustrates how the needs described by SDT align to *participant's theory*. While SDT only identifies three core needs, participant's theory adds two additional types of people: those who have their needs met and those who do not. It is important to understand these need

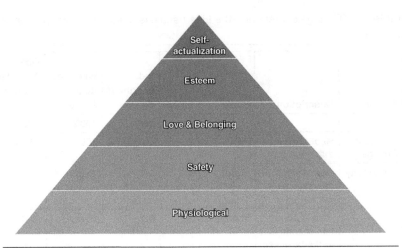

Figure 6.2   Maslow's hierarchy of human needs

Table 6.2   Types of change and the self-determination theory needs matrix

Change Type	Needs Threatened	Most Likely to React	
		Demographic	**Business Ecosystem**
Re-organization	Relatedness	Younger, Lower Seniority	Highly Social
	Autonomy	Older, Higher Seniority	Highly Individual, Performance Based
People + Culture	Relatedness	Older, Average Seniority, Mid-career	Highly Social, Performance Based
	Autonomy	Older, Average Seniority, Mid-career	Highly Individual, Performance Based
Process	Relatedness	All	Highly Social
	Autonomy	Older, Average Seniority, Mid-career	Performance Based, Long-term Stable
	Competence	Older	Performance Based, Long-term Stable
Technology	Competence	Older, Average Seniority, Mid-career	Highly Individual, Long-term Stable, Poor Leadership, High Internal Conflict

types and how they align, as well as the level of need fulfillment, so that the change agent can leverage this data to determine the most appropriate change tactics.

There is a strong correlation between the type of change being implemented, the dominant needs of the community, and the success

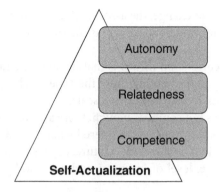

Figure 6.3   Self-determination theory and Maslow's hierarchy

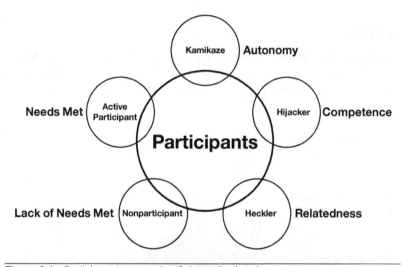

Figure 6.4   Participant types and self-determination theory

of transformation initiatives based upon the types of change tactics employed. Although, for the most part, people will react depending on their own strongest dominant need, people within a community or an organizational collective can reinforce the reactions of one another based on the type of culture within that community. This reinforcement creates a dominant need of the overall community or organization that becomes a part of the culture, or a *collective dominant need.*

Once the types of change are identified, it is important to understand the associated needs that relate to each type and then to identify the appropriate change tactics. Understanding that there is a direct relationship between the type of change, the associated needs, and the tactics that will best enable the team to be successful is the single-best method for overall change success.

What this correlation means is that people's reactions to change can be predicted by understanding how the type of change aligns to those three basic needs across the reaction scale.

As if this was not complex enough, there are layers of differing angles and perspectives that must be considered when planning change in order to make it holistic. These include: culture, capacity and capability, organizational structure, level of intimacy, and roles and responsibilities (see Figure 6.5).

In other words, change initiatives must be viewed from all of these perspectives in order to create a full and complete transformation.

Culture

Again, culture is the set of shared beliefs, practices, and behaviors that are considered socially acceptable or *normal* within a particular organization. It is a way of life for the people who work within the organization.

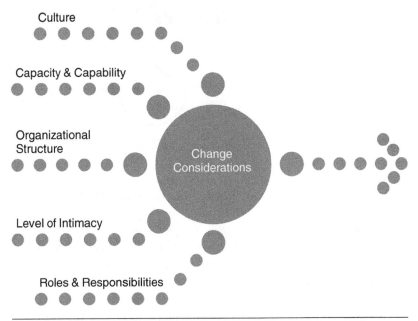

Figure 6.5   Change considerations

Culture is important to the discussion of change management because there is a direct correlation between the normal behaviors of everyday work life and how people will behave when it comes to organizational transformation. That is to say, bullies will try to bully their way out of it, manipulators will try to con the project team, while collaborators will find ways to contribute.

It is the role of the change agent to understand how people behave and interact with one another *before* initiating a transformational effort, in order to define the best strategies for managing the behaviors of the people impacted. This is because stress amplifies bad behaviors and brings out new ones—and change is stressful.

In terms of culture, it is important to identify exactly *what* is changing. Remember that culture is an amalgamation of aspirations, values, and behaviors. To change culture, you must identify all three elements and look for potential disparities between them. In this case, aspirations could be the same as the vision of the company, or they may be different for each executive. Ultimately, however, they do all have to align to the overall corporate vision, or the company culture will erode.

How Is the Culture Changing?

The next step is to determine how the elements of culture will change. This simply means to identify the discrepancies and create a strategy/plan for implementing those changes through reinforcement of key values and behaviors.

For Example

A leader may determine that he or she wants to have a team with less gossip and infighting, higher attendance, and higher overall productivity. This is the *what is changing*. The next step is to identify *how* this will be accomplished in a way that is in alignment with values and demonstrated behaviors.

To do this, an open and honest conversation with the leader about their values must take place so that those values can be identified. This way it can be made clear where the leader's values are not in sync with the behaviors being reinforced. To be clear, behaviors such as employee attendance and productivity cannot be adjusted without realigning the values of the leader because values have a way of being expressed through unconscious bias and behaviors that individuals may not recognize.

What Are the Impacts?

As with every other aspect of change management, it is important to identify the impacts of the proposed changes. This is especially so with culture because people form habits and a comfort zone about how to behave at work. Impacts can be identified with some basic questions such as:

- What are the aspirations of the leaders involved?
- What values do company leadership want to promote?
- What behaviors are changing?
- How will they change?
- What behaviors should replace them?
- What changes will be necessary as a result of these changes?
- Who is impacted by these changes?

Changing a toxic corporate culture is difficult because employees may not see the positive impacts on the horizon. Very often people who work in these environments are very insecure. They have to be because their job is constantly at risk. That means that all change, even positive changes that are in their own best interests, will be regarded with mistrust and trepidation.

That being said, it is also important to understand all of the potential ways in which people might react to the proposed changes, and then determine how best to communicate with them. This can best be done by talking with leadership and other stakeholders about past transformation initiatives to understand how people reacted and why.

Again, culture is the social environment—the fabric, if you will—of a given organization. People must be engaged and involved in transformational efforts from the start in order to ensure that they are fully prepared to adopt changes as they are implemented. This begins and ends with communication and engagement.

Communication is not a one-way transmission of information, but is a conversation between leadership and the general employee population. The types and levels of communication are important; not just because what is communicated and how is a demonstration of culture, but also because the emphasis on employee engagement is made clear in how leadership communicates.

Capacities

Capacity is the number of people on a given team with the skills and ability to perform specific functions. Those functions are referred to as capabilities. Sometimes, as the result of a change, the capacities of specific functions fluctuate up or down in order to meet the demands of the new service/product.

It is important to note that even when people consistently report being understaffed, overworked, and completely burned out, the addition of new resources can still create drama and trauma for the existing resources because they have worked hard to establish their social order within the company culture. New people threaten that in almost the same way that removing redundant resources from the work flow does. It means that people have to change relationships, alliances, and ways of working—and it means they need to effectively create a new culture.

Capabilities

Capability refers to the overall skills of the resources impacted by the change. Every transformation brings new skills that are related to the new tools and methods for performing the work. Organizations are a collective of these capabilities, and the overall ability to transform is highly dependent on the shift in skills at the individual level.

It's important to identify what capabilities are changing during the transformation process and to identify any transition skills. In other words, we have to identify the new skills required for the end state, but there may also be skills that are required for managing during the transformation that may be different from both the skills at the beginning and at the end.

Understanding how the capabilities are changing will enable the organization to fully prepare new capabilities for all the stages in between the start and end of the transformation. This includes knowing how the capabilities will be changed and over what time period.

It can be fairly simple to understand the impacts of changes to capabilities. What is important to note is how people adapt to new capabilities. And yes, they will take this personally. Remember our discussion about suddenly going blind? This is that scenario.

People initially react to changes in capability by erecting the unconscious fear barrier. This barrier keeps them from learning because they realize they don't have the new skills and they feel incompetent. When implementation and training efforts are initiated to help them learn the new capabilities, they gradually lose that barrier as they begin to feel more comfortable and competent with them. Once the transition is completed, they (should) feel completely competent.

The transformation in capability is not a short journey. Far too many companies poorly plan the training portion of implementation. They prepare a handful of sessions over a one- to two-month period. Then they wonder why people don't make more of an effort to attend—and accuse them of *not getting on board*.

First of all, without doing anything to address the barrier of fear and the feeling of incompetence, hosting a training session isn't going to attract people who cannot see past this wall to a time when they can be competent by following the learning plan. It is the role of change management to address that wall of fear and enable people to feel comfortable enough that they can and will attend the scheduled training events.

All of this takes trust. If the people impacted don't trust the project team, they won't listen to them and nothing the change agent says or does is going to encourage them to get on board or attend any training sessions. Earn their trust first. Then help them to get past their fear and then encourage them to attend the training.

Organizational Structure

Organizational structure is the framework for how the company is organized in order to achieve its goals and serve its customers. This structure often changes as a result of new leadership, changes in the economic situation, a new product or service offering, and mergers and acquisitions.

It is crucial to know exactly what parts of the organizational structure are changing, so that the appropriate strategies and plans can be executed. Some basic questions to help guide this discussion include:

- What parts of the organizational structure are changing?
  - One department or service line
  - An entire product division
  - The entire company
- Is it a merger?
- Is it an acquisition?

It is important to know these things because they represent the level of threat to an individual's job security and they will react accordingly. Therefore, the change agent must be able to identify and predict the impacts of change by identifying what elements of the organizational structure are changing and why.

It is just as important to know how the structure is changing as it is to know what and why it is changing because the way in which people react to changes in structure are highly dependent on these factors. Typical questions to pose would include:

- Are there more or less leadership roles being brought in?
- Are there more or less functional roles within the revised organization?
- What jobs are impacted?
    - Who currently fills those roles?
    - Is there a plan for those individuals to continue on with the company?
    - Are fewer people being asked to do more work?
- How is the authority structure changing as a result?

It is also important to understand the impacts of the organizational changes being made. Who is impacted—and by how much—are good indicators of how well people will receive the changes and adopt them.

Roles and Responsibilities

Roles and responsibilities are the job functions that people perform in their daily work routines, including those areas of authority that they hold over people and processes. When roles and responsibilities change, it is often because tasks, authorities, and specific functions are being added, changed, or entirely removed.

Think back to how people feel about their jobs. Remember that they usually integrate their professional role as a part of their personal identity. Changing roles and responsibilities can threaten that, and people do not usually just jump at the chance to change.

It goes back to inertia. But in some cases, it just seems easier to keep doing what they have always done, even if it is not necessarily aligned with the new vision of the organization.

PROCESS

Processes are those operational methods that are leveraged by the organization to get work done. In other words, it is how they do what they do. A simple process flow is shown in Figure 6.6.

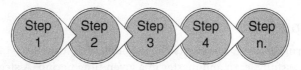

Figure 6.6   Process flow

Essentially, every organization is a collective of people who collaborate together by using specific processes and tools that support those processes. What that means is that people could wind up doing the same sets of things over and over again, all day long, every day—over the course of days, months, and even years.

The processes become second nature to them. They know how to get the job done without a lot of discomfort and fear. In fact, when people first meet others, often one of the first three things ever discussed is what they do for a living. It is personal, it becomes a part of their individual identity. So much so, that when processes change, people can feel personally threatened.

In terms of process, it is very important to define, understand, and spell out exactly what is changing, who is impacted, and how they can participate. Participation is key. When it comes to the personal nature of processes in the workplace, remember, many of these people have been doing these tasks for a long time. They've had time to think about how they would do it differently and what parts do not make sense and are not efficient.

Leverage their knowledge. Better yet, give them a chance to shine by incorporating their suggestions and ideas into the solution and give them credit for it. People support what they helped to build. They also like looking like heroes.

Types of Processes

The types of processes that are being changed matter a great deal because the level of process complexity, the amount of rigor and governance, and the number of integration points with other groups will increase the number of people who will feel the need to contribute to the changes. However, just because more people will feel entitled to make some contribution, does not necessarily mean that their perception is accurate. This, in and of itself, creates conflict.

Stakeholder perception, when coupled with an increase in stakeholders, effectively means that the change activities required for the project will have to focus heavily on collaboration, providing a sense of control, and training on the new processes.

Learning Curve

The learning curve represents the amount of information that people can successfully learn and master in a given time period. A steeper curve suggests that they can learn more material over a shorter time period;

while a longer curve suggests the opposite. The length of time that it takes to learn new processes is a factor worth considering when it comes to change, but so is the length of time it took to learn the old ones.

Why does that matter when they are probably dumping the old processes in favor of new ones? Commitment. The longer it took to learn and master, the more commitment people have to maintaining them, and the more resistance there will be to changing and removing them. The irony is that it is easier to get people to adopt new processes that are simple to master and have a steeper learning curve.

Demographics

When it comes to changing processes, the fact is that some demographics handle change better than others. Younger people cope with change a lot more readily than do older employees. All of this is important to know when developing the change strategy because activities, participation, and types of engagement will vary depending largely on the audience. That is perfectly fine. Know how they change, how they work, and how they prefer to participate—and there will be a lot less frustration and lot more change going on.

TECHNOLOGY

Knowing and understanding the specific details of any technology that is being implemented is as equally important as knowing the details for any other types of change. This is quite simply because there will be implications for the business areas and for how people perform their jobs. However, there is also another consideration for making changes to technology, and that is the demographics of the specific area impacted by the new technology.

Demographics

The demographics of the area impacted are especially important because seasoned employees are typically more comfortable with how things are done and are often more hesitant and resistant to technology changes. The point is that people who are older or who have been with the company for a long period of time can feel overwhelmed when it comes to learning new technology. When coupled with other types of change, the impacts are compounded.

Learning Curve

The learning curve on new technology can change dramatically based on the demographics of the stakeholders. A younger demographic is more accustomed to learning new technology and almost seem to be hard-wired to learn it more readily, while an older one may take time to adapt to the thinking and the rationale behind the technology before they can learn how to utilize it. It is important to understand that both demographics and the complexity of the technology will play an equally crucial role in the learning curve—plan accordingly.

7

Who Is Impacted by the Change

Mediation is based on the principle that you cannot control the actions and reactions of another person. However, it is also founded on the fundamental principle that by leveraging a specific process of self-control and self-disclosure through carefully crafted communication, you can mitigate the reactions of the other person. Mediation, therefore, seeks to promote understanding through open communication. Communication, according to this practice, is the two-way flow of information which enables all parties to understand and then to be understood.

To manage change means to control the flow of events, including communication (as discussed in Chapter 4). All too often, people in business ask: "How do I get buy-in for what we're doing?" The simplest answer to that question is that you don't. Why, you ask?

We don't seek to gain buy-in because that is really asking the wrong question. The message behind this question is really, "How do I get people to do what I want them to do?" Again, you don't.

Consider the goals of mediation, and let's ask the question differently. Instead of asking how you get buy-in, try asking, "How do we engage people in the changes being made so that we all benefit from this initiative?"

In order to determine who should be engaged in the changes, we must first determine who is impacted by the changes. To put that simply, there are three degrees of impact (as illustrated in Figure 7.1): direct, indirect, and no impact.

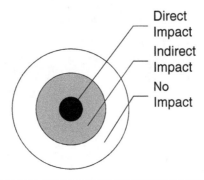

Figure 7.1   Degrees of impact

It is important to understand where stakeholders land in these degrees in order to understand:

- How engaged they need to be
- The types of participation needed from them for success
- The types of activities that can be leveraged to enable engagement
- The types of communications that can be leveraged to enable engagement
- The pace and flow of activities

It is therefore imperative that stakeholders be identified against the *degrees of impact* before needs and readiness assessments can be carried out.

STAKEHOLDER NEEDS ASSESSMENT

A stakeholder needs assessment (see Table 7.1) is a tool that enables the change agent to consciously and deliberately uncover information about the key stakeholders that will make the entire process of transformation dramatically easier. It does this by specifically identifying each stakeholder and their needs and allows the agent to understand the diversity among the stakeholders, identify potential conflicting needs, and establish a strategy that can address as many of the needs as possible. By leveraging this tool, the agent is able to make the best possible decisions about how to communicate change and how to engage the stakeholders for the maximum benefit to the organization and the overall process.

Table 7.1   Stakeholder needs assessment

Step 1: **Identify the** **stakeholders**	• What are the areas and business units impacted by this project? • What does each of these impacted areas do (what functions are they responsible for)? • What are the main concerns for this area? • How does each area fit into the organization? • How is each area going to be impacted? • How much is each area going to be impacted? • What will be impacted in each area? • How many people are in this impacted area? • Who are they?
Step 2: **Identify the** **needs of** **each** **stakeholder**	• What types of language do each stakeholder use? • What types of body language do each stakeholder use? • What does the problem to be solved look like for each stakeholder/group? • What kind of things does each group say about the problem? • What frustrates them the most about it? • What are their fears and concerns about a solution? • What should that solution look like? • Can you classify each stakeholder as one of: "Active", "Non-participant", "Heckler", "Hijacker", or "Kamikaze"? • What are the personal needs of each stakeholder in relation to their work environment? • What are their expectations of you or the project team? • Can those expectations be accommodated? • What are your expectations of each stakeholder? • What are the project expectations for the stakeholders? • Are there any differences? • Can differences be negotiated? • How can you manage/leverage these needs to support the project?
Step 3: Build **a trust bond** **with each** **stakeholder**	• Meet with each stakeholder individually when possible. • Ask each stakeholder to talk about the project from their perspective. • What are their success and failure criteria? • What are their personal concerns? • What does each stakeholder need to see for results? • Set up an informal communication plan to assure each stakeholder as the project progresses. • Follow through.
Step 4: Build **trust bridges** **between** **stakeholders** **and** **stakeholder** **groups**	• Are there any stakeholders and groups that are in conflict with each other about what is needed? • What is each group saying? • What are the similarities between each group's needs or what they are saying? • Talk to the groups in a combined meeting and point out similarities at every opportunity. • What are groups not in conflict saying? • Point out similarities in concern, needs, and interests between all groups. • Identify and talk about how each group will benefit from the new solution and how that aligns to overall business needs.

Table 7.1   (continued)

Step 5: Combine the needs in a single document	• Combine all identified business needs into a single document. • Create a mission and vision statement for the project (if not already done) from the results-based needs. • Write the vision and mission statements at the front of this document. • Post these statements in your workspace. • Distribute this document.

Egos in Decision Making

"Anytime there is a struggle between doing what is actually right and doing what seems right, then your ego is interfering with your decision."[1]

Change agents must always conduct a needs and stakeholder analysis for every transformation effort. The needs and stakeholder analysis is a trust-building activity that simply cannot be duplicated or replaced with any other task or activity, and failure to build trust results in a lack of adoption.

The needs and stakeholder analysis includes five basic steps:

- Identify the stakeholders
- Identify the stakeholder's needs
- Build a trust bond with each stakeholder
- Build trust bridges between the various stakeholder groups
- Combine the business needs of all stakeholders into a single document

The needs assessment is conducted by identifying your stakeholders and by meeting with each of them to identify their needs. It is important to note that there can be two types of stakeholders: visible and invisible. A visible stakeholder is someone who has an established purpose at the table during the conversations. An invisible stakeholder, on the other hand, is someone who is either directly or indirectly impacted by the changes being made, but has not been asked to the table by the business executive, even though they have the influence and authority to prevent others from adopting the changes.

Invisible Stakeholder

Recently, a mid-size insurance company was undertaking an enterprise-wide business transformation effort in order to change their market position and obtain a more favorable position against their competition. Within this company, a newly appointed vice president was not brought into the effort as a stakeholder, and was simply given the same input as many mid-level managers.

She disagreed with the transformation effort because she felt that it would negatively affect the business as a whole, despite the data and research to the contrary. As a result, the vice president was vocal with the people in her division about how she felt and about rejecting any of the changes being made by the project team.

In this case, this vice president was an invisible stakeholder. She actively lobbied the people in her division to protest and reject the changes being made. This was evident when a survey was distributed and results tallied by the executives since her division stood out as the most vehemently opposed to making the changes to improve the business.

In the aforementioned case of the invisible stakeholder, the change agent opted not to meet with the vice president and did not make any inroads toward getting her on board with the transformation. Unfortunately, this is all too common because people do not like to deal with unpleasant situations and discussions. However, that is exactly the reason that businesses *need* organizational change management.

Key Activities

The first task is to identify who the stakeholders of the transformation initiative are. In order to identify the stakeholders, the change agent must answer these basic questions:

- What are the areas and business units impacted by this project?
- What does each of these impacted areas do (what functions are they responsible for)?
- What are the main concerns for this area?
- How does each area fit into the organization?
- How is each area going to be impacted?
- How much is each area going to be impacted?
- What will be impacted in each area?
- How many people are in this impacted area?
- Who are the stakeholders?

Once the stakeholders have been identified, it is important to identify and understand their individual needs so that these can be incorporated into the project through appropriate engagement. It is imperative for change agents to remember that there is a difference between needs, wants, and expectations; and it is crucial to be able to identify each of them for the stakeholders in order to build a successful plan for involvement.

The differences between needs, wants, and expectations are not always apparent, so consider these statements:

- The stakeholder *needs* to see business results
- This same stakeholder *expects* the project to produce those results
- And, this stakeholder also *wants* the project team members to get along with the business team

While projects are initiated to meet business needs, the project team must recognize that they are dealing and working with people. It is the people who will carry the changes forward when the project closes, but they will only do that if the project has included them and treated them as if they are important.

In understanding the differences, the change agent can be more successful in supporting the transformation effort as a whole. To identify the stakeholder's needs, answer the following basic set of questions:

- What types of language does each stakeholder use?
- What types of body language does each stakeholder use?
- What does the problem to be solved look like for each stakeholder/group?
- What kinds of things does each group say about the problem?
- What frustrates the stakeholder the most about it?
- What are the stakeholder's fears and concerns about a solution?
- What should this solution look like?
- Can you classify each stakeholder as one of: *active, non-participant, heckler, hijacker,* or *kamikaze*?
- What are the personal needs of each stakeholder in relation to their work environment?
- What are the stakeholder's expectations of you/the project team?
- Can those expectations be accommodated?
- What are the business analysis team's expectations of each stakeholder?
- What are the project expectations for the stakeholders?
- Are there any differences?
- Can these differences be negotiated?
- How can the business analyst manage/leverage these needs to support the project?

Next, the change agent must build trust with each stakeholder. This is crucial but is often overlooked by project teams. Remember that the best stakeholder is an engaged participant because they will support the implementation and turn it into a grassroots movement.

However, there can be no engagement without involvement, and both of these require trust. The stakeholder will neither be engaged nor involved if they do not trust the change agent or the project team. In order to build trust, it is important to look for, understand, and perform the following:

- Meet with each stakeholder individually when possible
- Ask each stakeholder to talk about the project from their perspective
- Determine the stakeholder's success and failure criteria
- Determine the stakeholder's personal concerns
- Determine what each stakeholder needs to see for results
- Set up an informal communication plan to assure each stakeholder as the project progresses
- Above all else, *follow through*

The next thing to be done is to understand and to recognize what each stakeholder group brings to the project. Unfortunately, this is one of the places where conflict and office politics can rear their ugly heads. This can be the result of poorly managed mergers and acquisitions, interpersonal relationship breakdowns (friendships and advancement), poorly managed technology projects, or interdivisional competitions.

As a direct result of these conflicts and office politics, it is also crucial for the change agent to leverage their newfound trust with each of the stakeholders to increase team morale and build bridges between the stakeholder groups. These conflicts and politics can negatively impact the levels of engagement from the business, regardless of how much trust these stakeholders may have in the project team. To this end, conflicts must be managed to prevent issues from impacting the project. In order to build these bridges, the change agent must answer each of the following questions and perform the following tasks:

- Are there any stakeholders and groups who are in conflict with each other about what is needed?
- What is each group saying?
- What are the similarities between each group's needs or what they are saying?

- Talk to the groups in a combined meeting and point out similarities at every opportunity.
- What are groups not in conflict saying?
- Point out similarities in concern, needs, and interests between all groups.
- Identify and talk about how each group will benefit from the new solution and how this aligns to overall business needs.

Finally, the change agent must compile the business needs of all of the stakeholders into a single document. The following basic steps will guide the development of the needs analysis artifacts:

- Combine all identified business needs into a single document
- Create a mission and vision statement for the project (if not already done) from the results-based needs
- Write the vision and mission statement at the front of this document
- Post this statement in your workspace
- Distribute the final document

Once the needs and stakeholder analysis has been completed, the change agent must identify the impediments or barriers to the changes. This information is used to guide the transformation efforts over the course of the project.

STAKEHOLDER GROUPS

Stakeholders are often the most misunderstood group within the context of a project. From the project's purely functional perspective, the stakeholders are only those individuals and groups who have some influence or *stake* in the outcome of the project. This includes the sponsors and those providing input into the project in order to generate that outcome.

However, from a change perspective, everyone is a stakeholder. While not everyone can attend meetings and sit at the table to participate, individual stakeholders are nominated from the larger pool of people who are impacted to represent the stakeholders-at-large, or the stakeholder general populous.

This differentiation is important for change management because it represents the care and consideration afforded to the work to be done and how those changes will not only impact the general population of stakeholders, but also how they will respond to the transformational efforts. That being said, there are a few things to consider in selecting and working with stakeholders, such as: team structure; team culture and dynamics; team demographics; as well as team function, roles, and responsibilities.

Team Structure

Team structure is all about how the group is organized to achieve the services that they provide to the rest of the organization (their functional role within the organization). In short, it is who reports to whom within the team—and this arrangement can either be formal or informal.

It is important to consider team structure when selecting stakeholders because teams will have different styles of communication, as well as expectations and feelings of entitlement, based on how they work together and how they work within the organization. How teams are organized and function across the organization may differ greatly and this may cause a lot of angst when it changes or when there is a merger with another team.

Team Culture and Dynamics

This book has covered a lot of ground on culture already. However, that view has been from an overall organizational perspective, and not within individual teams.

Culture does not have a new definition because it is at the team level, but it most certainly can be different between the overall organization and the individual teams. Culture is heavily influenced by leadership, and as such, culture can vary dramatically from the organization to the individual teams that comprised it.

It is for just this reason that it is just as important to get to know how these teams feel, behave, and work. The transformation project may not be changing the culture of the organization as a whole, but it may in fact be changing one small corner of the world for a few people who feel very strongly about who they are and how they do things.

From a change perspective, the change agent should work to get to know how people will be impacted from each team's perspective—even if that impact is only the perception of something major. To be successful, it will be important to know and understand the nuances of the team culture of every team that is impacted by the changes, both directly and indirectly.

Team Demographics

Demographics are the statistical data of the team and particular groups (or cliques) within it. Effectively this is the age, background, and gender make-up of the team.

Just as it is important to understand the culture of individual teams within the organization as a whole, it is equally crucial to know and understand the demographics of each team. How people react is more than just impacted by how change is communicated, what is being changed, and the myriad of other factors discussed throughout this book. It is also heavily influenced by who the people on the team are as individuals. Remember that change, no matter how technical, is very personal to those expected to live through it and run with the new solution.

Team Function, Roles, and Responsibilities

This can be described as the purpose of the team within the organization. What primary function does the team perform—accounting, human resources, information technology, or sales?

The collective function, roles, and responsibilities of the team matter when it comes to change in the same way that they do for an individual. Teams, just like people, have a pecking order or a hierarchy within the organization that is based upon how important the function is perceived to be.

The sales team within a highly sales-driven organization may be seen as the most important group because they bring in the cash that keeps the business going. There can be feelings of entitlement that come along with this perception of importance, and these will most assuredly impact change.

The change agent should work to understand all of the nuances of teams and behaviors in order to determine how to best manage the changes and create a successful program of change.

STAKEHOLDER IMPACT ASSESSMENT

Unlike the needs assessment, the impact assessment describes how each stakeholder would be impacted and to what degree those impacts would be felt by each stakeholder and their group. In other words, just because change is coming, does not mean that it will impact everyone the same way and to the same degree. This is similar to the risk and impact assessment that project managers perform at the outset of a project.

As illustrated in Table 7.2, the impact assessment provides valuable insight into key areas that will need to be addressed in the strategy to successfully implement the changes.

Table 7.2   Sample stakeholder impact assessment

Stakeholder/Group	Likelihood of Impact	Degree of Impact
Accounting	High	Low
Human Resources	Low	Medium
Sales	Medium	High

REFERENCE

1. Johnson, Darren L., *Letting Go*, LettingGoCafe.com; 2nd edition; 2005.

8

OCM Approaches

AEROBIC VERSUS ANAEROBIC SYSTEMS

The human body contains two different types of systems for performing different types of work. These systems are called the aerobic and the anaerobic systems. The anaerobic system is designed to promote strength and speed and is utilized by the body for shorter spans of high intensity activities. The aerobic system, on the other hand, is designed for endurance over a longer period of time. Notably, both systems develop differently in athletes. Change, as with the aerobic and anaerobic systems, requires differing levels of endurance and effort depending upon the objectives of the transformation.

In managing change, it is important to know the type of initiative you are working on in order to determine the best approach and team selection. For example, you wouldn't choose a weight lifter to compete in a marathon, any more than you would equip a runner with the wrong shoes and training. To be clear, the approach is important. It is as important as selecting the right resources and tools for the change to happen.

We have already discussed that people resist change for many reasons and that they react to change very differently. It is for this reason that the selection of resources and tools is as meaningful as how change is approached within an organization. Therefore, it is important to know what approaches exist, how they relate to one another, and how to leverage them for the best result as a part of the overall change strategy.

ADKAR

The ADKAR model represents an approach to change management that depicts *change at an individual level*.[1] This model asserts that people change through a simple process: awareness, desire, knowledge, ability, and reinforcement. Effectively, it states that before people can change, they need to be *aware* that change is necessary. They must then have the *desire* to change; the *knowledge* of how to change; the *ability* to turn that knowledge into action and make the changes; and finally, the internal and external factors to sustain or to *reinforce* the changes.

The ADKAR model presents a concise framework for how to guide people through changes on small to medium projects where time is of the essence. In conjunction with other techniques and methods, the ADKAR model provides a solid strategy for managing large-scale transformational efforts as well.

The ADKAR model is really all about disseminating information to people about impending changes; convincing them to accept those changes; providing them with the training necessary to feel comfortable; helping people to see tangible success with the new methods; and providing the external support or performance system that ensures that they will largely comply.

Key Activities

Awareness

The primary activities prescribed for the awareness stage of the ADKAR model are all about communication and the distribution of information about the upcoming changes. The important factor here is that this information must be able to address the following questions from the perspective of those impacted most by change:

- What are the changes?
- Why are the changes being made now?
- What is wrong with the way we are doing it now?
- What are the risks of not changing?
- How will it impact our organization?
- What's in it for me (WIIFM)?

Desire

The desire stage is where the change agent ensures that people embrace the changes. In order to do this effectively, there are five tactics prescribed in this stage. These include:

- Leadership that sponsors the changes (executive sponsor)
- Enabling team leaders to become change leaders
- Employee engagement in the process of change
- Aligning existing incentive initiatives

Knowledge

The knowledge stage of the ADKAR model is focused on the training of people in the new skills that are required when the changes have been implemented. Training is the one area that ensures that people feel comfortable with the changes. This is because they will have the experience to enable them to more readily adopt those changes. The knowledge stage includes the following activities:

- Training programs
- Job aides
- Coaching
- User groups

Ability

In order to foster ability, the ADKAR model specifies activities that are designed to foster demonstrated capability in the new skills. According to the model, this includes:

- Daily involvement of supervisors
- Access to subject matter experts (SMEs)
- Performance management
- Workshop activities during training

Reinforcement

Finally, the reinforcement stage of the ADKAR model describes particular activities that ensure that the changes are made in such a way as to *stick*. This effectively means that the changes are made in an environment that encourages the people impacted to adopt them. These activities include:

- Making the changes meaningful to the individuals impacted
- Associating the changes with real and tangible accomplishments
- Absence of negative consequences
- Governance (accountability mechanisms) is in place to manage compliance

KOTTER: THE 8-STEP MODEL

The Kotter model, named for famed change management guru John Kotter, is an eight-step framework for leading change and transformational efforts. It originally was defined as a means of guiding leaders to drive change within their own organizations, but has since been updated to define the mechanism for accelerating change. The framework consists of the following steps:

1. Create a sense of urgency
2. Build a guiding coalition
3. Form strategic vision and initiatives
4. Enlist a volunteer army
5. Enable action by removing barriers
6. Generate short-term wins
7. Sustain acceleration
8. Institute change

The Kotter model provides a framework that is useful for the coordination of change activities across an end-to-end initiative.

Key Activities

Create a Sense of Urgency

The Kotter model defines step one as creating a sense of urgency—and describes this as the point in time when the change agent must communicate the so-called *burning platform*. Specifically, it elaborates that executive leaders must describe the opportunity for change in such a way that it will appeal to both logic and emotion. It is this statement that is leveraged to raise the needed army of volunteers.

Build a Guiding Coalition

The next step that Kotter advocates is to bring together a team with both authority and influence to lead the changes effectively, through the power of persuasion. He asserts that the volunteer army is more likely to follow a coalition of people from among their own ranks in order to guide and coordinate it, and to communicate its progress.

Form Strategic Vision and Initiatives

In addition, Kotter strongly recommends the creation of a compelling vision of the future to provide focus and guidance. This vision of the future is more likely to inspire people to want to change, but it will also help them to make the changes because they will know what the outcomes are supposed to look and feel like.

Enlist Volunteer Army

As mentioned earlier, the Kotter model urges change agents to identify and build an army of volunteers. This is the same as leveraging champions and evangelists who will advocate on behalf of the project simply because they believe in it. This is important because it ultimately turns the transformation into a grassroots movement.

Enable Action by Removing Barriers

Another important step for that change agent when utilizing the Kotter model is to work to remove the barriers to change, such as those discussed in Chapter 1. However, Kotter takes this a step further and suggests that this also involves removing inefficient processes and power structures.

Generate Short-term Wins

Next, it is crucial to plan high-profile quick wins that can showcase success with the changes in a short amount of time. This includes rewarding and recognizing those people involved in the execution of these quick wins, which will help to boost morale and get others involved.

Sustain Acceleration

Next, the Kotter model dictates that change agents must change those policies and procedures that are misaligned to the new vision, and then recruit and promote those employees who can implement the vision. In this way, the model asserts that changes are more likely to be sustained because those tasked with keeping them going have been rewarded and promoted for doing so.

Institute Change

Finally, the model advocates for making the changes by calling out and identifying the connections between success and the new ways of working. In other words, help people to redefine success so that the new definition they rely on and look to includes the new processes, tools, and structures being implemented. In this way, people will actively work toward the achievement of success by the new definition.

IIEMO

IIEMO is an acronym that stands for: inform, involve, evolve, maintain, and observe. It is an approach that was developed out of conflict resolution and mediation theories as a means of engaging and collaborating with people through change, as opposed to having change thrust upon them by leadership or external forces beyond their control.

The IIEMO approach is useful in that it focuses on the consultation and collaboration between leadership and the people who are not only impacted by change, but also are tasked with implementing and adopting it. As such, each stage is characterized by increasing communication and collaborative activities that enable people to get involved, learn and change together, and above all, to trust in the changes and leadership while going through the process.

When those impacted by the changes are left completely out of the communication loop, as is far too often the case, they typically feel ignored and unimportant which causes them to lose trust in leadership. In fact, when you hear statements such as, "I don't know, I just do as I'm told," people are expressing their feelings of disenfranchisement and lack of involvement in the process of change.

This approach contains stages with various goals, activities, and tasks for change managers to apply and work through that will increase the likelihood of getting people to collaborate. Above all else, the primary task is to communicate change to those impacted and to give them opportunities to come to the table and collaborate in a safe environment.

Again, the IIEMO approach is leveraged in stages that enable a gradual build-up to acceptance of change. The philosophy is that people change more readily when they are allowed time to digest and understand the changes rather than having it thrown at them. The following text goes into further explanation of the IIEMO stages.

Inform

The inform stage is characterized by the extent, level, and quality of information disseminated about the upcoming changes. To this end, the basic task is to formulate the informational packages and distribute them across multiple channels of the communication architecture, including: e-mail, meetings, intranet, internet, bulletin boards, and notices.

Goal

The overall goal of the inform stage is to ensure that as many people as possible are informed that change is coming. There are two things to keep in mind when it comes to informing people: give them as much notice as possible and make sure the information is as accurate as possible.

The last thing that any organization can afford is rumor and innuendo from unannounced changes; or loss of credibility from too many unimplemented change announcements. Unannounced changes do more than make people feel unimportant to the organization—they make people feel nervous about job security. Unimplemented changes, on the other hand, teach people not to trust or believe in leadership. In either case, the organization ends up with an attrition problem, and most certainly will have a much more difficult time implementing changes that stick.

Audience

The typical audience for the first stage includes: executives, employees, customers, vendors, partners, and shareholders. Much of this is to disseminate the information that specific changes are coming and to prepare them for more specific details about the changes and who will be impacted.

Involve

The involve stage is executed in two phases. It consists of planning for preliminary meetings and information sessions. The phases are simply called phase one and phase two, and each has a similar objective that relates back to the overall stage goal.

Goal

The goal of this stage is to involve stakeholders in waves and to achieve buy-in from the right people at the right time in order to support the activities and encourage a smooth transition of the project from a leadership perspective. Usually, workshops within this session are appropriate and highly recommended.

Phase I

Audience

The typical audience for this first wave of participants in phase one includes: managers, team leaders, and program and project managers who are responsible for ensuring the buy-in, adoption, and ongoing compliance of their functional and project teams.

Expected Participation

Everyone who is involved in the activities and sessions of this phase will be asked to contribute to the discussions, including discussions as to who is impacted by the change and how, as well as to make recommendations for improvements and implementation.

Change Agent Session Task

During sessions held with this wave of people, the change agent is tasked with identifying those who seem resistant, obstinate, or against the change. They also need to identify those people who have additional questions, then arrange to spend time with them in small groups or as individuals.

These people will be the champions and the ambassadors for change. It is critical that they understand and buy in to the changes in order for this to happen.

Phase II

Audience

The audience for this next wave of participants in phase two includes: employees and members of impacted teams.

Expected Participation

The expected participation is similar to that of phase one; everyone involved in the activities and sessions of this phase will be asked to contribute to the discussions and to make recommendations for improvements and implementation. However, there is one slight difference. In phase two, there is also a learning element for participants to get information about the tools and processes that are changing and to realize why they are important and how they can be leveraged to help them perform their roles more effectively.

Change Agent Session Task

The job of the change agent in phase two is the same as it is in phase one: identify people who appear to be against the changes, along with those who simply have additional questions, and arrange to spend time with them. Again, these people will be the advocates for change when dealing with their colleagues, customers, and vendors. Never underestimate the power of one negative voice in the crowd.

Evolve

Goal

The objective of the evolve stage is to complete the requisite changes prescribed by the transformation initiative. Effectively, it is the *action* stage where the organization sets about implementing the changes that it has decided are necessary and has been talking about. This may include building the software, revising the business processes, creating the new program, moving to the new location, opening the new store, launching the new product, and the list goes on and on.

What is important to note is that communication does not stop during this activity simply because the team is now hard at work. In fact, communication must increase during this stage. Why? Because nothing makes people more nervous and uncomfortable than a lack of communication! This stage is critical for ongoing engagement so that the end result is something that the people impacted can leverage or work with.

Audience

The audience for this stage is anyone that will be impacted by the changes, anyone considered SMEs, anyone who is a key stakeholder, and anyone who is working as a sponsor of the transformation.

Expected Participation

Each person identified in the audience has a different role to play in the evolve stage, as they do in the execution of the project. Their contributions to the success of the transformation initiative are crucial to extending the changes outward to the entire organization.

The people impacted are asked to contribute time and ideas for the development of the changes, as well as to test the results in order to ensure that the changes are effective. In order to accomplish this, sessions can be held for brainstorming and identification of requirements for the changes.

SMEs are asked to contribute their depth of knowledge and expertise in key areas in order to refine the concepts into solid changes that can be implemented to improve the ongoing operations of the organization. To accomplish this, sessions are held to elicit knowledge and to run through processes with them in order to define better requirements.

Key stakeholders are asked to contribute organizational knowledge about the big picture and objectives that help to direct the efforts of the SMEs and the people impacted. Stakeholders are kept abreast of the work efforts being done and are asked to contribute to making decisions about the overall requirements for the changes.

Finally, sponsors are the senior executives that sign off on the transformation from a high level and can support the team by demonstrating commitment to change and removing roadblocks. They ensure that the transformational efforts will deliver on the business need and are accountable for managing the overall budget and initiative by directing the project manager. It is imperative to keep the sponsors informed about key transformation activities and how well things are being received by those impacted; as well as to escalate critical issues when necessary.

Change Agent Tasks

During this stage, the key tasks of the change agent are to ensure the full engagement of everyone who is required to contribute to the project. To this end, the change agent will be reviewing schedules, recommending additional sessions, and coordinating communications.

Who Must Be Engaged?

It is a simple task to determine who has to be engaged in changes throughout the evolve stage—*everyone*. That is to say that everyone who is invited to any project meeting for determining scope, requirements, testing, and development of the product must be engaged. Remember, engagement is what drives people to contribute, and a lack of engagement will turn into a lack of adoption in the end.

First and foremost, the agent will review activities, events, and meetings scheduled by the project team to determine the level of engagement required by the participants. If there is a gap in the engagement, they will make recommendations to increase engagement and will schedule additional events and activities to ensure full coverage of all participants in the engagement throughout this stage.

Finally, the change agent will be coordinating communications between the project team and the people impacted in order to ensure that information is disseminated from the team at an appropriate rate, and that feedback is given to the team. In other words, the change agent is facilitating communication between the two sides in order to ensure that engagement is maximized for the best possible results.

Maintain

Maintaining change is really about sustaining the momentum across the entire initiative in order to ensure that full adoption occurs, thus resulting in a complete transformation of the organization. As such, there really is no maintaining stage per se, but an overarching attention to the establishment of key metrics and milestones for success and progress during the development and implementation of the new system. This occurs both at the outset, in the planning of the changes, and during the process of execution of the plans.

These key metrics and milestones are useful across the process of transformation in that they allow the organization to better understand the people impacted by change, and they also allow for the change agent to plan and execute the change strategy in a way that is meaningful to those impacted. It is crucial that the evolution process of the change be closely monitored to ensure adoption, allowing it to be tweaked where and when results are deviating. This helps to maintain alignment between the people impacted and the changes being made.

Goal

The goal of the maintain stage is really to ensure that the project team can sustain the momentum of change throughout the entire transformation process. Far too often, interest in the transformation wanes and people become disengaged throughout the process.

There are many reasons that this can happen. These include:

- The project takes too long to deliver
- People don't see enough of a difference when they expect it
- People are not actively involved in the process of change
- The project team does not incorporate suggestions and lessons learned from the people impacted
- Other projects in the past have failed
- People don't know how to be successful in the new way of doing things
- Implementation of the changes occurs in an illogical order

Because of these issues with consistent involvement and engagement, it is important to ensure that there is a way to measure the adoption rate and the success of the transformation initiative while it is underway, so that tweaks can be made to realign those impacted by the changes.

Change Agent Tasks

In order to maintain the momentum of the transformation, the change agent must take some time during the planning of the project to establish the metrics and milestones that will be leveraged to monitor the health of the overall execution—and to determine how well the changes were implemented. Knowing what can be measured, how, and why is crucial to making these decisions. Of course, other factors will impact the decisions about what to measure—the types of changes being made, the business climate, and the environment will all influence the outcome of this task.

Observe

The observe stage is where the change agent actively monitors the process of change. It is much like the maintain stage, but is more of an overarching and ongoing task that covers the span of the project. In effect, it is the task of monitoring the people and the process of change by leveraging the established metrics to measure the success of the transformation as it is progressing.

This task is important and useful when change agents remember that people are not simply a rubric, but are the ones who will carry a change forward. By observing people as they go through the transformation, the change agent can spot areas of concern and prevent disengagement—ensuring a higher adoption rate in the end.

The observe stage is about leveraging metrics to understand underlying issues and areas where people may not feel safe to open up and talk, but will definitely show a level of disengagement in the adoption rates and the metrics being measured. Above all, this stage is designed to ensure that the change agent remembers that a large part of their role is to act as the concierge of change by providing a degree of personalized service to groups of people.

Does that mean that a couple of change agents can provide individualized and personal service to thousands of people impacted by change? No, but it does mean they could and should be providing service that feels personalized by leveraging the techniques and the information about groups of people in order to create a careful change strategy.

The task of observing is conducted by taking routine measurements of the established metrics against the milestones and by interacting with groups of people to understand who they are and what their unique needs are.

Goal

The goal of the observe stage is to ensure alignment of the changes made to the behavior demonstrated by those on the front lines through governance. Since the best way to understand what has changed and by how much, it is important to have milestones in place to observe and measure the changes thus far.

Change Agent Tasks

The main tasks of the change agent during this stage are to monitor the progress of the changes, measure the differences against the baseline, and report those differences to project leadership, stakeholders, and sponsors. That being said, it is advisable (as will be discussed in Chapter 12 on gamification) to ensure the main stats are also reported back to those impacted by the changes.

Monitor

Monitoring the changes is simple. The fact is that once the changes have been planned, there is a large part of the change agent's time that is dedicated to observing and watching how people are reacting to the changes.

Change Agents and Anesthesiology

Consider the change agent to be like an anesthesiologist during surgery. They may not be doing any of the cutting, but they sure do make it easier to persevere without a lot of unnecessary pain. Just like the anesthesiologist who doesn't just put the patient to sleep and leave the room, but monitors and watches the vitals of the patient until it's all over, the change agent should keep watch over the change process.

Measure

Many of the measures taken as part of the baseline in the change scorecard can be remeasured at various stages throughout the transformation effort in order to gauge just how much progress is actually being made. Think of it like earned value management or the burn-down chart for change.

While the IIEMO approach does not specifically call out or identify what to measure, it does rely on common sense. Questions such as, "How much has changed?"; "How fast did it change?"; and "Did it really change?" can be answered by measuring things such as the adoption rate and the sentiment change ratio.

Report

The change agent should have a reporting method to inform people about the progress of the transformation. It is both a means of informing and motivating.

Charity Giving

Recently, many companies held a week of activities designed to raise money for charities such as the United Way and the Red Cross. Within each of those companies, posters were put up and e-mails were distributed regularly to inform the participants where each group and the company as a whole was in relation to the goal. Every time the amount raised was increased, the posters were changed to reflect the progress.

As a means to inform, reporting helps to support and enable reaching the objective through tactic changes and shifting scheduling and activities. Project leadership is able to leverage this information to estimate the actual implementation dates as well as the true cost of adoption based on these changes in scheduling.

As a means to motivate, when people can see the progress they are making, it often inspires them to work harder to achieve the objective. As in the case with gamification—when these same people see how well their group fares against another on the same track toward the goal, it often encourages them to compete and work to achieve the goal faster than the other teams.

It is important to understand that the IIEMO approach is founded on a logical planning process and that it is supported and enabled by the observe stage. Realistically, this stage underpins all other stages and ensures that the tactics, activities, and methods leveraged to effect change, are in fact working. The observe stage provides the specific information about what is working and what is not. It is this particular knowledge that enables the success of the transformation efforts.

AIDA

In the world of marketing, AIDA is an acronym that stands for the basic formula for copy in order to encourage sales prospects to buy quickly. The acronym stands for:

- Attention
- Interest
- Desire
- Action

So what is a marketing acronym doing in a book about organizational change management? Consider this: change management is really a big sales job. It is the job of convincing people to stop what they are already doing and to do something else. Realistically, that is the objective behind change management.

Table 8.1 illustrates the alignment between the AIDA principle, the ADKAR approach, the Kotter model, and the IIEMO process.

The AIDA principle helps to define key tasks and activities that make change easier, in the same way that every one of the approaches does. It provides guidance to practitioners and change agents to help gain buy-in from those directly impacted by change.

The key difference here is where the other principles and approaches focus on how people change (i.e., what do they need to change), the AIDA principle is focused on how to convince them that the new solution, process, or situation is going to be better. AIDA focuses on showing people how the future looks with the new changes implemented in the same way that advertising shows people how great life would be with that shiny new washing machine.

Attention

The whole objective of this stage is to get the attention of those directly impacted by change. In this case, they are the consumer audience. Various techniques can be employed here such as:

- Direct e-mail campaigns
- Town hall announcements
- Notifications on websites (both internal and external)
- Webinars
- Posters
- Staff meetings

Table 8.1   Alignment between AIDA, ADKAR, Kotter, and IIEMO

ADKAR	Kotter 8-Step Model	IIEMO	AIDA
Awareness	Create a sense of urgency	Inform	Attention
	Build guiding coalition		
Desire	Form strategic vision and initiatives		Interest
Knowledge	Enlist volunteer army	Involve	
	Enable action by removing barriers		Desire
	Generate short-term wins	Evolve	Action
Reinforcement	Institute Change		
	Sustain Acceleration	Maintain	
		Observe	

Interest

The interest stage is used to engage the audience on a deeper level so that they will want to spend their valuable time understanding the key messages of the transformation effort in more detail. Obtaining the audience's interest is more involved than simply grabbing their attention.

The audience will give the change agent a little more time to build interest, but that agent must stay focused on the needs of the audience. This means helping them to select the messages that are relevant to them very quickly. As a tip, when creating marketing and informational materials for the transformation effort, use bullets and subheadings to break up the text to make important points stand out.

Desire

The desire stage, on the other hand, is about helping those impacted by change to really want those changes to happen. This is where the storytelling and the visioning come in. It is important to help people understand how life will look, feel, and be once the transformation has occurred. It's like showing them how much the solution can actually help them.

The best way of showing them this is by appealing to their personal needs and wants. As mentioned earlier, change—all change—is personal. It is crucial, therefore, to focus messaging on those needs and wants.

Instead of saying things like, "This session will teach you how to use the new software," explain to the audience what's in it for them: "Get what you need from other people, as well as save time and frustration, by learning how to use the new tool."

Features and Benefits

Another good way of building people's desire for the change is to leverage the link between features and benefits of the new solution.

When it comes to the marketing copy for the transformation efforts, it's important to remember the benefits. When describing the changes, do not just give the facts and features of the new solution and expect the audience to realize the benefits for themselves. Be explicit and clearly tell them the benefits related to each feature in order to create that interest and desire.

> **Example**
>
> Feature: "We are implementing new firewalls and policies."
>
> Audience: "So what?" or "Who cares?"
>
> Instead, persuade the audience by adding the benefits:
>
> Benefit: "We're making it even easier for you to access the websites you need every day."
>
> You may want to take this further by appealing to people's deeper drives: "This may result in not having to work overtime."

Action

Finally, the action stage is where the change agent must be very clear about what action people must perform. Here again, the agent must not leave people to figure out what to do for themselves because they either will not do it or may end up doing the wrong thing.

> **Example**
>
> Attend the workshop on Friday to learn everything you need to know to succeed.
>
> Register now.
>
> Update your laptop no later than Friday at 4 PM.
>
> Take the training online today.

REFERENCE

1. Hiatt, Jeffrey M.; *ADKAR: A Model for Change in Business, Government and Our Community*; Prosci Learning Center Publications; 2006.

9

Organizational Change Manager Techniques

There are a lot of creative things that a change agent can do to ensure that a transformation is a resounding success. These change techniques may work in one organization, but not in another. It is imperative for that agent to have a toolkit at their disposal in order to fully engage as many stakeholders as possible.

Just as people learn in many different ways, people also process information in different ways—and at different rates. The change agent can capitalize on this when selecting the best techniques to employ in any given change scenario.

> People learn more, and learn faster when they think they're not even learning at all.

VISIONING

Vision Statement

A vision statement is a short descriptive sentence that identifies the overall image or objective of what is being built. In other words, what the organization will look like once the transformation has occurred and is completed.

The vision statement for change initiatives is critical because it provides people with something to work toward. It effectively enables people to keep their *eye on the ball.*

When people are working in the *trenches* so to speak, and they get bogged down in the seemingly endless stream of minute tasks that comprise the whole effort, they can often lose sight of direction—and this leads to poor decision making. This is exactly how so many projects end up off course and delivering something other than what was promised.

Having a vision statement, however, means that there is always a lighthouse to aim for and a yardstick by which to measure and align decisions. When people have a vision statement, the team can ask, "How does this get us to our goal?" This type of decision making is more likely to keep the project on track.

The vision statement should contain prophetic language about what the organization will look and feel like in the future state. A good example of a business vision statement is this one from comScore: "To leverage the power of the Internet to increase the effectiveness and efficiency of our clients' sales and marketing efforts."[1]

But what about smaller objectives within a single organization that are really a part of helping them to achieve their overall vision? Are they the same? Well, no. They are different in that the statement for a single change or transformation effort focuses on the results of the particular change that is being made. A good example of this would be this one that was written for an enterprise-wide transformation to a company's information technology (IT) infrastructure: "Create an integrated IT infrastructure and delivery framework that supports the business to its *full* potential."

Mission Statement

Many people often get vision statements confused with mission and unique service proposition statements. A solid mission statement should describe how the change or transformation will achieve the vision.

Mission statements for individual change efforts are intended to unite the project team, the stakeholders, and any others who may be directly or indirectly impacted. Like vision statements, mission statements that are written for change efforts enable people to focus on the activities that must be completed. However, in this case, those activities are to achieve the vision.

In the case of an enterprise-wide transformation effort, there may be a single overarching vision statement and multiple mission statements to correlate with the multiple projects that are required to achieve the end result.

Both vision and mission statements are recommended for large-scale transformation initiatives that consist of multiple projects; however, only one is recommended for single, one-off projects. Again, the value of the mission statement is to unify and enable the team to focus not only its own attention, but that of the people who are participating, as well as those impacted.

Mission statements contain clear language about what is being worked on. A good example of an organizational mission statement is this one from Amnesty International: "To undertake research and action focused on preventing and ending grave abuses of these rights."

COMMUNICATION ARCHITECTURE

Communication architecture, as illustrated in Figure 9.1, is a framework for the planning and management of communication activities on a project in order to ensure that communication is explicit rather than implicit. The framework includes plans, policies, and tools. It specifies the key messaging and the tools by which communication will occur.

Communication architecture is the infrastructure that is leveraged for communicating throughout the transformation effort. Organizations make the assumption when they hire and bring on resources that these

Communication Architecture

Plans	Artifacts	Tools	Policies
• Communication plan • Escalation plan • Crisis plan • Communication schedule	• Risks and issues log • Key messaging	• Infrastructure • Email • FAQ pages • Surveys • Chat rooms	• Point of contact • Response times

Figure 9.1 Communication architecture

resources know how to communicate in a professional manner and that they are following a specific set of invisible guidelines.

The trouble is that every organization has invisible guidelines—differing invisible guidelines—and that they are, well, invisible. What complicates matters is that these guidelines change depending on the personal preferences of the people involved. This sets people up for failure. It is no wonder that one of the most common causes of project failure is the lack of engagement.

Communication architecture takes the implicit invisible guidelines around communication and makes them explicit so that everyone can follow them.

Benefit of Communication Architecture

Communication architecture transforms implicit expectations for communication into explicit methods and policies for professional conversations.

In terms of change and transformational efforts, communication architecture is the critical framework that will be leveraged to disseminate information out to those people who will be impacted. It will allow the change agent to collect feedback from them in such a way that they can feel empowered and important to the process.

As shown in Table 9.1, communication architecture contains very specific elements that can be leveraged in combination with one another to enhance the levels of communication within the organization.

Table 9.1 Communication architecture

Element	Item List	Definition
	RASCI matrix	Defines roles and responsibilities of team members, sponsors, and stakeholders
	Lessons learned log	Outlines lessons learned during the process of executing the project that would improve the way the company executes future initiatives
	Risks and issues log	Tracks and monitors the risks and issues identified during the execution *and* to be addressed by the project
Artifacts	Key messaging	Specific messages to be sent to stakeholder groups, project team members, and user groups at various points during project execution

Element	Item List	Definition
	Vision and mission statements	The vision statement articulates *what* the company will look like once the solution has been implemented. What are we building? Think VISIONARY. The mission statement defines how the vision will be achieved. Think MISSIONARY.
Plans	Communication plan	Outlines the types, audience, and frequencies of specific internal and external project communications.
	Escalation plan	Outlines the plan of action to be taken, including who is to be informed when execution is not aligned to timelines, budgets, and resources.
	Organizational change management plan	Outlines the plan for marketing the new solution, engaging the business stakeholders and users, business readiness, and solution training.
Tools	Communication infrastructure	Defines static tools for records, documents, and knowledge management.
	Input funnels	Defines the specific tools to be utilized to collect input from the stakeholders and user groups, including e-mail accounts, social media, chat rooms, surveys, etc.
	Outgoing channels	Defines the specific tools to be utilized to distribute messages from the project team to the stakeholders and user groups, including e-mail accounts, social media, FAQ pages, chat rooms, etc.
Policies	Point of contact	Identify the primary and secondary points of contact. Consider who will be allowed to disseminate information outside of the project.
	Tool usage statements	How and when will the architecture tools be utilized in the project in order to help it achieve its purpose?
	Critical messaging	Identify the critical messages that might be necessary to distribute from the project to the employees and/or customers; knowing this in advance helps reduce time to respond.

Artifacts

Artifacts are those tangible work products that enable the change agent to better manage the communications on the transformation project. While these work products are helpful on the project and for future transformational efforts within the organization, the primary purpose of this documentation is to support the coordination of the change activities.

RACI Matrix

As briefly mentioned in the introduction, a RACI matrix is used to define the roles and responsibilities of the team members, sponsors, and stakeholders within the overall project. However, this artifact is particularly helpful to the change agent for determining the primary contacts for activities and engagement of teams throughout the transformation.

The RACI matrix contains a listing of the key stakeholders, their position on the project, and their relationship to the project. These relationships, as shown in Table 9.2, include:

- Responsible—indicates the person is responsible for specific tasks on the project
- Accountable—indicates that the person is accountable for ensuring that those responsible complete their specified tasks
- Consulted—indicates that the person is to be consulted on decisions and will contribute to key decision making without actually participating in executing the associated tasks
- Informed—indicates that the person is merely to be informed of progress or completion of the key activities

For the purposes of change planning and coordination, the change agent may opt to create a change-specific RACI matrix describing who among the stakeholders, subject matter experts (SMEs), and sponsors is responsible, accountable, consulted, and informed of change tasks and activities.

Table 9.2   RACI matrix

Name	Position	R	A	C	I
	Project manager		X		
	Testing SME			X	
	Interface SME			X	
	Business unit SME			X	
	Business unit SME			X	
	Business unit SME			X	
	Business unit SME			X	
	Business analyst	X			
	Business unit manager			X	
	Business unit manager		X		
	QA lead			X	
	Development lead				X

R–Responsible, A–Accountable, C–Consulted, I–Informed

This will enable the change agent to better plan for the transformation and result in a more successful execution of those plans.

Lessons Learned Log

The lessons learned log is another artifact that is designed to support the overall project effort by recording the crucial lessons that have been learned during the process of execution that would improve the way that the company plans and implements future initiatives. The lessons learned are intended to serve as a record for issues to be addressed *after* the project has ended.

Every team that participates in the planning, delivery, and implementation is responsible for creating and maintaining an individual log, as well as contributing to the overall project log. Change management is no exception.

Within the context of change, the lessons learned log is an invaluable source of information about the types of activities that increased motivation and led to higher success and adoption rates, as well as those that decreased them. In highly contentious transformations, this log can be a rich source of information as the project progresses.

As illustrated in Table 9.3, the lessons learned log contains an overall category title to describe the lesson, a specific issue name, a basic description of the problem or the success, a brief overview of the impact, the recommendation, and the identity of who (by role or by group) is accountable for escalation or implementation of the recommendation.

Risks and Issues Log

Similar to the lessons learned log, the risks and issues log tracks and monitors the risks and issues identified during execution. In this case, however, these items are to be addressed by the project in order to complete that same execution.

Again, this log is intended to be contributed to by every team and every team member. Each of the teams (analysis, delivery, testing, and change) should also generate and maintain an independent log for their specific efforts and roll this information up to the project level when appropriate.

In terms of change management, the risks and issues log enables the ongoing management of those risks and issues that are specifically related to the ability to implement the changes or to execute the transformation.

Table 9.4 depicts the basic contents of the log. It includes: risk class (critical, medium, and low), risk name, risk description, mitigation strategy, accountability, probability, risk manager, occurrence, addressed, and resolution.

Table 9.3   Lessons learned log

Category	Issue Name	Problem/Success	Impact	Recommendation	Accountable
Procurement management	Contract requirements	The PM was not fully engaged in the contract process.	All requirements were not included in the initial contract award. A contract modification was required that added a week to the project.	The PM must be fully engaged in all contract processes. This must be communicated to both the PM and contract personnel.	Manager of procurement
Human resources management	Award plan	There was no plan for providing awards and recognition to team members.	Toward the end of the project, morale was low among the project team. There was increased conflict and team members were asking to leave the project.	The PM should institute and communicate an awards/recognition program for every project.	Human resources manager

Table 9.4   Risks and issues log

Risk Class (Critical, Medium, Low)	Risk Name	Risk Description	Mitigation Strategy	Accountability	Probability (%)	Risk Manager	Occurrence (Y/N)	Addressed (Y/N)	Resolution

Key Messaging

Key messaging is the specific messages (copy) to be sent to stakeholder groups, project team members, and user groups at various points during project execution. These messages are crafted to announce changes, schedules related to those changes, and specific details to inform those people who are impacted about what to expect.

Having these key messages crafted during the planning stage, to the greatest extent possible, prevents change agents from reacting to e-mails, phone calls, and other inquiries without due consideration for what is being said. It is far too easy to respond to angry e-mails with knee-jerk negative responses that do nothing but breed contempt and divide the teams and the business. This does nothing to support the successful transformation and is exacerbated when the responding team member is ill-informed or releases information at inappropriate times.

The key messaging contains carefully crafted statements from project leadership and the organizational executive. These messages are intended to be delivered at specific intervals for the purposes of:

- Announcing impending changes
- Discussing the need for change
- Distribution of the vision and mission statements
- Responding to inquiries about the transformation
- Automated response e-mails from the project accounts
- Distribution of scheduling and training information
- Surveys
- Requests for input
- Meeting notices

Vision and Mission Statements

The vision and mission statements, as previously discussed in Chapter 5, articulate the end goals of the transformation once all is said and done. Specifically, the vision statement articulates what the company will look like once the solution has been implemented, and the mission statement defines how that vision will be achieved.

In the context of change, these statements help to tell the people within the organization a story of where they will be when they are successful and this transformation is long behind them. This type of story is motivating and provides the guidance needed to help people actively shape that vision in everything they do by aligning to the new standards.

Plans

Plans are those deliverables that enable the successful execution of the transformation by laying out the clear path for each of the tasks to be completed—who is involved, when things will be accomplished, and the estimated costs associated with those tasks. No activity, event, or task related to change should ever be unplanned. As they say, not planning is the same as planning to fail.

Organizational Change Management Plan

The organizational change management (OCM) plan details the overall plan for marketing the new solution—engaging the business stakeholders and users; business readiness and solution training; governance of the methods to be utilized; as well as the resourcing and estimated costs for the transformational efforts.

Far too often, change management is an afterthought by stakeholders and sponsors who do not understand that people within the organization require coaching, guidance, and support through any kind of transformation. That need only increases as the effort gets larger and envelops more and more of the organization.

The change management plan enables the team to demonstrate tangible value to the project team, the organizational executive, and the stakeholder community by creating a solid plan backed by governance and metrics that can prove the effectiveness of the events and activities undertaken as part of the change initiative.

This plan, similar to a project plan, contains resourcing, costs, activities, scheduling, goals, and scope. It firmly draws the box around the transformational activities, how to prove the changes are successful, and how to monitor the progress of those efforts. This enables the change agents to take corrective actions when the changes are not going as planned.

Communication Plan

The communication plan is similar to the project-level plan in that it outlines the types of internal and external communications planned for the transformation effort. It is leveraged as a tool for planning and enables transparency between the change team and other groups, such as the project teams at large, the stakeholder groups, and the greater organization.

In addition to planning the communication types necessary for the change initiative, it also identifies the purpose, who is responsible for delivering it, the audience and format for each of these types, as well as the frequency of distribution. Table 9.5 illustrates a commonly used plan format.

For Example:

An internal frequently asked questions (FAQ) page is planned for the big move to new office space. The communication plan identifies:

- Communication type: Static communication
- Communication purpose: To provide a place for people to find answers to common questions about the move
- Delivered by: Sandra Beacon
- Audience: ABC Company, internal only
- Communication format: Intranet FAQ page
- Frequency: Updated weekly, as required

Escalation Plan

An escalation plan outlines the course of action to be taken in the event of an urgent situation during the transformation. This plan includes who is to be informed when these situations arise. This is not to be confused with the standard change control process that is activated when technical issues arise during an implementation.

Unplanned Situation

A number of years ago, an insurance company was implementing a program to penalize drivers for specific offences since these particular driving habits were often seen in the records of people with higher accident claims. Just as the holidays hit and the company was in slowdown for two weeks, the project team sent a letter to customers who would be impacted.

That letter caused a lot of confusion and heated emotions. This led to an unexpected (unplanned) volume in calls to the call center by these customers wanting to talk about this letter. The call center was inundated and overrun with calls that it did not have the staff to manage.

In the scenario just described, the change management team did not have a plan in place to manage this kind of urgent situation, and it got very ugly, very fast. Even if it is never utilized, an escalation plan is always advisable for transformation efforts, especially when those efforts include people who are outside of the organization.

The escalation plan, as described above, is useful in planning out who will be contacted in the event of key situations arising on the project. Having this plan in place will ensure that the response to these situations, should any arise, is well coordinated and managed. This response will

Table 9.5 Communications plan

Communication Type	Communication Purpose	Delivered By	Audience	Communication Format	Frequency
Identify the general descriptive title of the communication. See examples below.	Describe the purpose of the communication.	Who on the team is responsible for delivering the communication?	Who will receive the communication?	Which format type is the communication delivered in? (There may be more than one type.) Use the project documentation checklist to identify if a specific template should be used.	How frequently is this type of communication required? List any specific days/times if known.
Examples					
Status updates	Inform of status of project activities	Project team	Project manager	E-mail, use the weekly status report template	Weekly, Friday, 12:00 P.M.
Review	Discuss current progress and set weekly goals	Project manager	Project team	Meeting	Weekly, Monday, 9:00 A.M.
Status reports	Inform of status of project activities	Project manager	Client project manager, project sponsors	E-mail, report	Weekly, Monday, 12:00 P.M.
In-process reviews (IPRs)	Inform of status of project activities, provide updates to work plan, and provide performance reports	Project manager	Steering committee	PowerPoint presentation, report	Monthly, first Tuesday of the month, 10:00 A.M.
Steering committee meetings	Discuss issues and changes affecting project outcomes	Project manager and selected team members	All	PowerPoint presentation, report	Semi-monthly, second and fourth Tuesday, 10:00 A.M.
Quality review(s)	Provide objective review of projects to ensure adherence to policies, processes, standards, and plans	Quality assurance manager	Project manager	Report, meeting, 1-on-1	Quarterly

salvage some of the lost credibility and will result in more people accepting the changes.

The escalation plan document contains the following information that will be critical to the execution of the plan:

- Escalation RACI
- Who to contact, when, and how
- Who to contact if others are unavailable
- What information to provide
- Executive phone tree
- Contact list (including cell phones and e-mail addresses)

Crisis Communication Plan (optional)

The crisis communication plan is very similar to the escalation plan in terms of some of the details, however this plan is only to be activated in the event of a crisis. The key difference between them is that the escalation plan describes who to contact and how when things do not go as planned and a significant problem is created; and the crisis plan describes what to do and where to go when an emergency evolves; for instance, if the building catches fire (hence the name *crisis*).

This type of plan is very helpful when the project team, including the change management resources, are outsourced; the organization does not already have one in place; or the transformation project is doing something that could potentially cause a crisis. This plan designates an alternate location as a communication office and includes distinct infrastructure for carrying out communication activities.

The crisis communication plan contains elements to support and enable the quick reaction to a crisis. These elements are as follows:

- Alternate communication office
- Separate communication infrastructure (network, phones, radios, televisions, etc.)
- Points of contact
- Designated people permitted to issue press releases or to speak to the press
- Dictates where personnel should go (physical location), when, and why
- What those personnel are expected to do once they are at the specified location
- Executive phone tree
- Contact list (including cell phones and e-mail addresses)

- Scheduling of resources for 24/7 management of communications until the crisis has passed

Tools

Communication Infrastructure

Communication infrastructure includes the physical tools that are required for communication throughout the duration of the transformation project. This can include everything from e-mail accounts to SharePoint project documentation sites.

This infrastructure is helpful on every project, but even more so when large-scale changes are being implemented. Here again, however, this is often an assumption made by the project team about what tools are currently being employed by the organization, and to some degree, that of any outsourcing partners.

Generally, project managers and teams make the assumption that e-mail will be utilized to communicate with others about the work being done; they make the assumption that if the particular organization uses a common document repository, that it will also be used here; and they make the assumption that in all likelihood, e-mail will be the most widely used form of communication. A question that should be asked is: is this what is best for *this* particular transformation effort?

The purpose behind identifying and specifying the types of communication infrastructure that will be leveraged across the effort is to change the way that people think about communication and to deploy the best tools possible for achieving the transformation goals. If the objective of communication architecture is to take implicit conversations and make them explicit, it is important to utilize the communication infrastructure that, when coupled with key messaging, will most support this.

Ad hoc Infrastructure

A number of years ago, one team lead was leaving a multi-year project and transitioning the responsibilities for his role onto a new team member. The new member was not only new to the project, but was also new to the company.

Unfortunately, the outgoing team lead communicated everything in e-mail format—and so did the team and the stakeholders. This meant that there were issues, risks, requirements, suggestions, documents, and everything you can think of that was related to the project—more specifically, changes to the scope, schedule, and budget—hidden in e-mails.

In order to successfully transition the new team lead and turn over the responsibilities to her, the outgoing lead simply spent days forwarding batches of e-mails from the previous two years so that the new leader could *dig out* whatever she needed.

In the previous scenario, it's easy to see that a few work logs and project management tools would have helped immensely. However, those weren't planned, and the lead was free to choose however he wanted to manage. It just was not really effective because upon review there were dozens of risks, issues, and suggestions that were missed because they were not tracked properly.

The truth is that communication is implicit in the business world. People are taught to read and write, and to do a job with specific skills. The assumption, then, is that they have enough skill to communicate well and professionally when it matters the most. This is just not an accurate assumption.

This infrastructure, as a part of communication architecture, is intended to provide the necessary guidelines for communicating specific information at appropriate times and in ways that make it easy to manage the transformation.

Incidentally, when logs and infrastructure are well used and managed, people actually feel heard. So change becomes that much easier and a lot less chaotic.

Communication infrastructure specifically identifies particular tools that will be employed across the transformation effort. These can generally be classified in one of two ways: input funnels or outgoing channels. However, that being said, there are other static tools that may be leveraged for the transformation that do not fall into one of these two categories. These include tools for records, documents, and knowledge management.

Input Funnels

Input funnels are the overall classification for the specific tools that will be utilized to collect information from multiple sources, such as project team members, stakeholders, and user groups across the entire project (not just change management). These tools are identified early on so that communication with the entire project team is seamless, and people can focus on doing the work without having to also navigate confusing invisible expectations or make assumptions about how to get their information to the right people in the right format.

When the analysts have a predetermined method for people to send in feedback or requirements, it makes it easier to locate and track them. The same can be said for risks and issues being sent in to the development or testing teams. One of the single most common issues with requirements is that they are often embedded within e-mails, documents, and conversations. Then stakeholders wonder why their need was ignored.

Input funnels include: project-related e-mail accounts, social media accounts, chat rooms, surveys, working sessions, and even focus groups. This does not mean that these tools cannot also be classified as either events or outgoing channels. Tools can be leveraged to facilitate the communication with the team and from the team.

Outgoing Channels

Outgoing channels are the complement to input funnels in that they define the specific tools to be utilized by the project team to distribute messages about the transformation to the sponsors, stakeholders, and user groups.

As with the input funnels, identifying particular outgoing channels before the project is initiated helps to create a platform for consistent communication with those who are impacted by change. When communication is consistent, people more readily participate and contribute because they have an idea of what to expect and then proceed to communicate in ways that are aligned to the styles and formats that make the contents of those communiques more readily consumable. This is a two-way street.

Outgoing communication channels are those tools which enable and facilitate the communication from the project team outward. These tools include: project-related e-mail accounts, social media accounts, FAQ pages, chat rooms, etc.

Policies

Point of Contact

The point of contact is made up of the key people who will be the face of the project when connecting with stakeholders and user groups to field comments, complaints, and information requests from the team. This person does not replace the analyst, the project manager, or others who may work with individual users and user groups to develop, test, and implement the solution. The point of contact is generally more than one person—primary and secondary.

This role is particularly helpful in alleviating stress from other team members being asked the same questions over and over again, and ensures that a single point of contact is available to mediate discussions that may become heated, or already have become heated, about contentious issues. Consider also how helpful it is to have a single point of contact serve as an intermediary between business people and techies. Let's face it, they are speaking two completely different languages—English and Geek-speak.

I Don't Care What You Think

Several years ago, a company was implementing a new software solution for customers and customer service agents as a self-serve portal. The lead developer was a wonderful person who had great ideas and logical rationale for why her suggestions should be implemented.

Unfortunately, the developer was not very good at articulating her suggestions in ways that the business could openly discuss them. It led to a lot of disagreements with the business every time they had a conversation. It left both sides feeling unimportant and unheard.

While the aforementioned scenario is a fairly mild example of some of the situations that occur, it wasn't until an intermediary stepped in and worked with both the lead and the business that they both realized each side had some valid points and great ideas. It made the solution that much better when, finally, one was not trying to steamroll the other, and both sides ultimately felt heard.

It is important to consider who will be allowed to disseminate information outside of the project; not only because of situations like the one cited, but because having a single point of contact creates consistent messaging and increases the level of trust that people will have in the project team and the organization.

As mentioned, this is very simple. Identify the roles and responsibilities of the point of contact, and identify at least a primary and a secondary person who will be tasked with communicating outward, especially in tense situations where it is important to present a unified front and deliver a consistent message.

Tool Usage Statements

In terms of policies, it is important to craft statements that describe how and when the architecture tools will be utilized on the project in order to help it achieve its purpose. These statements will describe each of the tools that have been identified in the infrastructure and describe any expectations around using them that might be considered *invisible expectations*. This is not the same as an appropriate usage statement.

Document Repositories

A company uses SharePoint as its primary document repository for projects. Each project is provided with a site to store all related artifacts and deliverables that each team member creates and works on.
  The SharePoint usage statement is as follows:

  Each team member is required to store all working and finished copies of their documents in the folder; and to separate them into subfolders by type (plans, processes, requirements, test cases, etc.). Each document must follow a specific path to sign off (the workflow), and each must have a copy of the approval e-mail stored in the same folder. This e-mail is distinct, and uses very specific verbiage to request and to provide approval. There can only be one document approved in any single e-mail.

The above scenario is an example of the kind of usage statement that would be leveraged to help provide guidance around the policies for using SharePoint within that organization. It is then crystal clear what is expected of each team member. Having this statement crafted in advance and available for all team members will establish consistent use of the tools across all resources and will shorten the time necessary to bring any new members up-to-date.

Critical Messaging

Unlike key messages which are delivered at routine intervals for regular project notifications, critical messages are those messages that are distributed to the project team, stakeholders, and sponsors in the event of a non-routine situation. Some of these non-routine situations can include prolonged scheduling delays, dramatic increases in budget, a need to reduce employees, problems with implementation, activating mitigation strategies for worst-case scenarios, etc.

Knowing in advance what to say and when, not only helps to reduce time to respond, but it also prevents knee-jerk reactions that could divulge too much information, create additional risk, and decrease the overall trust that people have in both the project and the organization.

NEEDS ASSESSMENT

As discussed in Chapter 8, the stakeholder needs assessment enables the change agent to discover information about the personal and professional preferences, needs, and goals that will make the entire process of transformation smoother. Again, this is more than an attempt to find out obscure information that provides leverage in negotiating, it is a trust-building activity that directly impacts development and implementation of the solution as well as overall adoption.

IMPACT ASSESSMENT

In Chapter 8, the stakeholder impact assessment was defined as a task that analyzes how and to what degree each stakeholder and their groups would be impacted by the changes. Understanding this crucial information effects change in that it enables the change agent to focus on those groups that are most affected by key aspects of the solution in order to gain the appropriate perspective in the development of the solution.

People need to have a feeling of control over change because they want to feel that they have control over what happens to them. The change agent and the project team are responsible for facilitating more than just input into the solution, they also have some degree of control over it with the stakeholders. It is this task that supports the decisions about what control is appropriate to give, to whom, and when. Ultimately, it increases the buy-in and engagement of the stakeholder community in the development and adoption of the solution.

Managing to the Exception

A friend of mine tells a great story about a company that wanted to reward its employees so management sat in a room to brainstorm about how they would do this. After spending several exhausting hours sequestered in the room, the most popular suggestion was to have management waiting by the elevators one day to offer free coffee to each employee as they arrived.

However, one of the managers objected to this idea and protested vehemently. She cited that this would not appeal to everyone, so they continued to go around in circles trying to come up with other ideas. This went on for another several hours, and everyone was getting tired. They kept coming back to the same idea. Eventually, in her exasperation, she protested that there must be approximately two people out of several hundred that drank tea instead of coffee.

They served coffee.

In this story, inappropriate control was given to stakeholders without knowing and understanding the full impacts of the solution. Had they known, the conversation could have been much shorter.

In change, many people have this idea that running kaizen events is a lot like this. They take too long to run because it is difficult to get everyone to agree. But the truth is, not everyone has to agree for change to be successful.

They do have to be willing to work together, and they have to be encouraged to participate in constructive ways by enlisting them to participate when and where it is appropriate for them to do so. There are many other ways to help people feel heard and included. That, after all, is really what change management is all about.

ROAD MAPPING

In its simplest definition, a road map is a type of plan that aligns both short-term and long-term goals with specific solutions that will help to achieve those goals. "It is a plan that applies to a new product or process, or to an emerging technology."[2] A change road map, however, is specific to identifying the transformation of behaviors, attitudes, and beliefs of the people impacted by change. Without this transformation, true change will be limited at best.

Road mapping is an important technique for change management because it provides context for the tasks and activities being worked on, but it also provides people with a very solid map of how they will get to where they are going. All of this is important, especially once people are in the thick of the transformation, forget about the vision, and lose sight of the direction.

Road mapping is conducted across three phases: preliminary, development, and follow-up. Each of these phases leverages specific sets of steps, artifacts, and deliverables in order to create the road map. By leveraging these steps, artifacts, and deliverables, the team is able to produce a road map that illustrates the alignment between activities and outcomes.

Key Activities

In the preliminary phase, key decision makers must be able to identify a specific problem and determine if road mapping would be useful in solving it. The key activities for each of the phases are described as follows:

Preliminary

- Satisfy the essential conditions for the transformation to occur (what groups must be involved, how they should be involved, and what will facilitate/encourage/promote their involvement)

- Provide unbiased change leadership and sponsorship of not only the transformational effort, but also of the road mapping process
- Define the scope and boundaries for the transformation road map that outlines the scope of the transformation as a predetermined set of needs, a planning horizon, and an appropriate level of detail

Development

If we follow the path laid out by the Defense Logistics Agency, but instead, apply a change management set of activities, it would look like this:

- Identify the specific change(s) that the road map will focus on
- Identify the critical transformation (behaviors, attitudes, and beliefs) requirements and who will be impacted by the transformation
- Specify the major change areas
- Specify the business drivers
- Identify the transformational alternatives and timelines
- Recommend the transformation alternatives to be leveraged
- Create the transformation road map

Follow-Up

Finally, the follow-up phase involves validation and sign-off of the transformation road map by the change team, as well as by the key stakeholders. Throughout the transformation effort, the road map must be reviewed and updated because the needs of the people impacted are evolving as the changes are being implemented.

KAIZEN EVENTS

Kaizen is a Japanese term that simply means *to change for the better*. In practice and application, it has come to signify continuous improvement. A kaizen event, on the other hand, is any action where the output is intended to improve upon an existing process.

In terms of change, the incremental nature of the changes being implemented makes the overall transformation seem much less daunting and scary. That, in and of itself, makes change easier to accomplish and implement.

In addition, kaizen events are also useful for convincing people and organizations of the value of the changes being made. However, the real intent of a kaizen event is to hold small activities that are attended by the key stakeholders involved in order to make improvements to the processes that they utilize every day.

Toyota is a good example of an implementation of the kaizen principle. Line personnel are expected to halt production lines when an abnormality appears and collaborate with their supervisor to suggest improvements that would resolve the abnormality. This may or may not initiate a kaizen event following the plan, do, check, act process.

According to Robert Maurer, Ph.D., "Kaizen is the art of making great and lasting change through small, steady increments."[3] In transformational efforts, this philosophy and process can be leveraged to enable smaller changes that enable the evolution of the organization in a more organic way, rather than artificially.

This type of change effectively makes everyone responsible for how well the organization is run, and commits everyone to the objectives. But it also ensures that people have more buy-in to the work that they do. Above all, when run effectively, kaizen events can make change easier.

It is important to note, however, that kaizen is *not* the same as managing to the exception. Kaizen events focus on the collaboration of people who perform the work to improve how it is done, while managing to the exception is allowing the small minority to dictate to the larger group based on their own personal preferences. Therefore, kaizen events are not always appropriate, but are highly dependent upon the maturity of the people involved.

SKUNK WORKS

Skunk works is similar to a kaizen event, however, organizations proactively encourage people to participate in groups to try out new methods, tools, and products based upon interest and not simply the work they currently perform. In effect, skunk works is a type of research and development lab that is funded by the organization to consider and test out the ideas of people within the organization.

Again, when people are involved in the generation of ideas that come to fruition and impact the way that the company does business, the products it sells, and how it interacts with its customer base, they feel a sense of ownership. This encourages and creates an environment of change where people tend to be more dynamic, involved, engaged, and participatory, as well as open. All of this makes change and transformation organic in nature and is embraced by the people impacted.

Typically skunk works is run as a form of research and development with a team of people that have a high degree of autonomy and are not restricted by the usual red tape, policy, and procedure. This team is often responsible for working on advanced or secret projects.

It is this type of environment that encourages people to think outside of the box to find solutions to problems and develop interesting new products. All of this encourages a high degree of engagement and promotes a change-ready atmosphere.

HACKATHONS

Similar to skunk works, a hackathon is a collection of people that are brought together to resolve problems. The key difference here is that a hackathon can be utilized to identify and resolve problems that people see every day on the job. In a way, a hackathon is a combination of a kaizen event and skunk works.

Hackathons are great places for people to come together and collectively solve problems and advance the business through creative working groups. What it does for team work and camaraderie is phenomenal. What it does for change readiness and the willingness to embrace it is stellar.

When people help to build something, they will support it. But if they helped to develop the idea and then bring it to fruition, they own and nurture it. They are empowered, and when it comes to other people's ideas and suggestions, they become more engaged in helping them bring their ideas to life as well.

All of this makes for an environment that is constantly evolving and changing. More than that, it teems with life, respect, and willingness to change.

Hackathons are typically run as team building events at several points throughout the year, as many times as an organization is willing to sponsor them. People from all different roles and groups come together and explore ideas that they would like to resolve, and they form teams to create an idea and pitch it to the rest of the team, who then decides which ideas will be developed.

Key Activities

Session Objective Discussion

First and foremost, it is important to set some focus for the event by having a discussion forum to talk about the objectives for the session, as well as the ground rules and the expected participation. This opening not only serves to help people focus on the event, but it helps to set the tone and provide a much-needed kick off.

Group Formation

Once the kick off has been initiated, participants are asked to break into groups for the duration of the event. This task is fairly simple, however, it is imperative that people who may be absent at this time to also be afforded the opportunity to join a group so ensure that people not present during this stage are assigned to a group.

Brainstorming

Brainstorming has two segments: problem identification and solution proposal. The objective of problem identification is simple: identify a problem within the organization that must be solved and obtain consensus from the group about the problem—there will likely be more than one passion area. In other words, everyone has to agree that this is a problem and that it needs to be resolved. The objective of the solution proposal is to brainstorm ideas for resolving that problem and then to agree on only one solution. The most important thing to remember here is that everyone in the group must be heard.

Planning

Once the team has identified their problem and a potential solution, they set about planning what that solution would look like and how they are going to demonstrate it to the entire group. Specifically, decisions are made about prototyping and assigning development roles. The most important thing to remember here is that everyone in the group must be included.

Development

It is during the development stage that the prototypes are designed and built. These prototypes will be leveraged for demonstration purposes during the presentation stage. The most important thing to remember here is that everyone in the group must be included.

Presentation Planning

Planning the presentation involves deciding how to present the problem statement and the proposed solution to the other groups. However, this is also the stage where the presentation materials are developed. This could be PowerPoint slides, skits, songs, logos, solution branding, posters, etc.— the most important thing to remember is to make it catchy.

Presentations

Next, the presentations are conducted in front of all of the other groups, and they are asked to vote on the best ideas. The most selected ideas are then voted on by a larger audience, or often just leadership. The most important thing to remember here is that everyone in the group must be included in the presentation, but in a way that makes them feel appreciated for their skill or talent.

Panel Voting

Finally, panel voting is where the larger audience (such as everyone in the organization) or leadership, is asked to vote on their choice for the top three ideas. Typically, this provides recognition for the best ideas and encourages everyone. The top idea, however, is usually the one that actually gets the go-ahead for further development and refinement.

QUICK WINS

As discussed in Chapter 4, quick wins are a type of mini-project such as a pilot or proof of concept that applies the new processes and procedures that are being changed. The key thing to note here is that quick wins are intended to be *quick*. In other words, they are intended to be short bursts of effort that demonstrate immediate results at a low cost. The quick win is intended to showcase the changes being implemented before they are finalized. In some cases, it can be called a soft roll out.

This is primarily done as a way to refine the end result, but it can have some serious implications for change adoption and organizational transformation. Remember that people need to see how something will work and know how it will feel to use it before they can feel comfortable and confident in leveraging it and adopting the changes. By applying quick wins, people have an opportunity to do just that. Moreover, when they start seeing success with it, they are more likely to share that success with their colleagues and become champions of the changes.

Since quick wins are effectively small pilot projects that will test the finished *product* (system, application, processes, etc.), they contain all of the elements that are set to be implemented, as well as a way to review and feed the information back into the transformation process.

To be clear, it is not the change agent's role to create and execute the quick win. It is, however, their role to participate in the planning and to monitor and observe the execution in order to ensure that it supports the development of people's confidence in the new method of operations.

That being said, what change agents need to know about the quick win is that it is conducted in stages: planning, project/event selection, execution, and review.

Key Activities

Planning

Planning quick wins is really the establishment of the objectives, success criteria, milestones, flow of work, and how the results will be leveraged in the bigger initiative. It is the determination of how many quick wins will be conducted, how they will be measured, and how that information will be utilized.

This plan will enable the team to maximize the effectiveness of the quick wins, monitor the progress, and then apply what has been learned back into the initiative. However, from a purely change management perspective, this plan will enable the change agent to measure the levels of confidence and buy-in of the teams during the execution, and it will enable them to ensure high levels of trust are maintained throughout a well-executed event.

Project/Event Selection

The plan objectives for the quick wins are the single-best means of determining the ideal projects and/or events to be leveraged. By identifying how people should feel and what they should know at the conclusion of a quick win, the change agent is able to suggest projects with similar goals and objectives.

First off, since quick wins are intended to be short bursts of effort to support the initiative and showcase the results, typically, the projects selected should be no longer than three months in length, and events should be no longer than a few days. In some cases, training can be leveraged with activities to supplement the quick wins.

Revised IT Security Protocols

Recently, an international insurance company underwent an enterprise-wide transformation of its IT security protocols. At the conclusion of this transformation initiative, the company ran a week-long, hands-on training session where the new processes were walked through in a classroom setting.

This provided the opportunity for practitioners to learn the processes and see how they could benefit the company by applying them to real scenarios in a stress-free environment. It served as a quick win because it enabled the people impacted to gain confidence in how the processes would be executed, while simultaneously testing them.

Next, an ideal project for a quick win is one that will utilize as many of the major new processes and systems as possible. When considering a test plan, a test manager plans for and analyzes test coverage to ensure that the bulk of the functionality is going to be tested. This ensures that they can confidently say the system works as planned.

Execution

The execution of the selected quick-win projects is where the real test of the changes occurs. In effect, it is applying the new processes, tools, or techniques to the project as if it were a regular project, with one exception: everything about this execution revolves around adapting the changes to a real-world environment and learning from how the solution is ultimately applied.

Review

Reviewing the quick wins is where the results and the lessons-learned logs are analyzed and discussed to see how the real-world application of the solution differed from the design—and how to realign the two. This is done by either changing the application or by changing the solution. This ensures that what has been designed is a practical and pragmatic solution that is useful and can be readily applied by many teams.

DETERMINING THE LEARNING PATH

Determining the learning path for people impacted by change is about planning the activities required to help people adopt the changes. It really is a process of learning and experience that enables people to more readily adopt the newly implemented tools and protocol. This includes planning for specific activities, literature, workshops, training sessions, focus groups, and team-building activities.

The best way to accomplish this is to understand the destination, plan the trip, identify the values you want people to espouse, and work with individual groups to find out how they learn and collaborate best. By leveraging this information, the change agent can plan the best path to learning—one that takes the impacted stakeholders into consideration.

One of the most critical success factors in change is determining how to effectively train people in the methods and skills necessary to fully implement the new systems, tools, and processes. Yet far too many change agents consider this to be as simple as a transfer of knowledge. This is more about making people feel comfortable and confident with the changes and making them feel competent when doing the work of applying them.

This is precisely why this is called determining the learning path, as opposed to training or knowledge transfer. The terms *training* and *knowledge transfer* imply the dissemination of information; whereas the *learning path* is all about the process of helping people develop the skill and confidence to perform.

Key Activities

Understand the Destination

Every well-planned journey starts with an understanding of the destination. Where are you going? It is the role of the change agent to be able to understand exactly where the organization wants to be at the end of the transformation. This is where the vision statement comes in, but it is also about the look and feel of the organization and the culture that is desired.

Far too many change efforts and projects focus on the academics and the bottom line of the organization without considering the *people* who will actually live and work within it. The destination is more about the shift in mind-set among the people that enables the change in the financial situation, in the academics of the organization, and in how it is run.

Plan the Trip

When planning the learning path, it is important to understand and to leverage various adult learning techniques, as these will support the development of new skills within the participants. The focus on these techniques is to instill the greatest amount of skill in the learners within the shortest amount of time—because time is money in the corporate world. As a result, these programs are concentrated on hands-on experiences and open discussion on the topics to help cement the skills along with the potential variations of each, as well as the logic behind them.

Identify the Values You Want People to Espouse

In planning a learning experience, it is important to consider the values that you want to impart to the learners as a part of the training process. Kids are often told by their parents and other adults to *do as you are told* and *do as I say, not as I do*. This attitude does not work very well when dealing with adult relationships in organizations.

As change agents, it is important to identify the new attitudes, mindsets, and values that will make up the transformed culture so that these can be infused into the learning. Ultimately, this will support the transformation of the people who are impacted by change into the organization as envisioned by the leadership.

Work with Individual Groups to Find Out How They Learn and Collaborate Best

---

Requirements 101

A few years ago, a retail company wanted to train several of its newly appointed business analysts in how to do the job they had been assigned. These people were struggling because none of them had ever worked in technology in the past, and each of them had previously been customer service representatives.

The person who was brought in to plan and conduct the training realized very quickly that the first barrier to overcome was the fear of failure that was brought on by their complete lack of exposure to anything related to IT. One woman even cried at the trainer's desk, explaining how incompetent she felt in the role.

Instead of diving right in to teach the people how to collect requirements, the trainer instead opted to have a short cooking course that was designed to show each of these people that they had transferrable skills from everyday life that they could use.

The essence of being a change agent and facilitating the development of new skills—whether they run the training themselves or not—is to ensure that the training is appropriate for the learning audience. This is the single-best way to ensure that people actively participate and that the training is appropriate for them—given not only their current skill set, but also their current frame of mind and attitude.

What happens when people are trained at a level they have already mastered (or at least believe they have) is that they become disengaged. They start saying things like: "Well that was a waste of time," or "That was eight hours I'll never get back." This sours not only *their* attitude, but the attitudes of those around them. Suddenly, people are not showing up for training, and adoption becomes an uphill battle.

CELEBRATING SUCCESSES

Celebrating the successes of the project team in making incremental changes as it progresses (such as at the end of a quick-win project) helps to motivate people to work harder and achieve more. It goes hand in hand with the gamification of change management as discussed in Chapter 12.

While these celebrations do not have to be big and expensive, or even large-scale, they do have to be sincere and meaningful. Creativity is always helpful and appreciated when planning these celebrations.

Custom-made Chocolates

A project team for a large consulting company was executing a project for a government client. Since government employees are not exactly known for their work perks, the project manager decided to have custom-made chocolates with the project logo molded into them.

Each little chocolate bar was individually wrapped in gold foil and handed out when the team and the stakeholders achieved a milestone. It went a long way toward helping people get involved in meetings and engaged to achieve milestones.

SPECIFIC COMMUNICATION CAMPAIGNS

Specific communication campaigns are marketing-style campaigns that disseminate information about the transformation effort. They are leveraged to inform and educate people about what to expect when they are approached to get involved, to inspire them to volunteer to champion the changes, and to help motivate them to adopt the changes.

Communication architecture is specifically developed to support these campaigns as a means of communicating with stakeholders and user groups. These campaigns are a valuable way to reach out to and to engage stakeholders and users on their own terms and in their own comfort zone.

Scenario A

A retail company spent a great deal of time and money to develop and implement a new project methodology. It took a couple of years to achieve it, and they were confident that they would start making great strides in executing successful projects.

Unfortunately, use of the new methodology declined quickly and did not achieve the results that it promised. Why? Most people simply did not even know about it.

In retrospect, what happened was two-fold: inadequate training opportunities and a lack of internal marketing. The project team had simply not run any campaigns to inform people about what was being implemented and how it would benefit them.

Contrast the preceding scenario with the following success story:

Scenario B

A local transportation company was in the process of upgrading its website to include the ability for passengers to purchase tickets online. They held a press conference to announce the changes to their customers and periodically sent out updates to let stakeholders and customers know about the progress of the project.

When the new website went live, online sales exceeded projections, and page views were exceptional.

In Scenario B, the customers were ready for and educated about the changes. They readily adopted the ability to purchase tickets online and to book passage without having to wait in long lines at the departure gate.

These campaigns must include a variety of mediums in order to be successful. Not everyone will open that e-mail, pass by that poster, or visit the company intranet site, but by leveraging multiple formats, the change agent increases the likelihood that people will find out about and talk to others about the transformation. This, in and of itself, will increase the chances of people finding out about the changes and then mentally preparing for them.

Key Activities

The key activities are the same as for any commercial advertising initiative.

- Define the purpose of the campaign—why is it being run?
- Identify the target audience—who is it for?
- Determine the message—what will they be told?
- Determine the medium—how will it be distributed?
- Determine the schedule—when will the campaign start?
- Determine the measures—how will success be measured?

NETWORKS AND PRACTICE COMMUNITIES

A community of practice is a common network that is similar to an association but is exclusively within a single organization. The people within the community share a skill set and actively work to enhance and advance that skill set. Some organizations also call this a *center of excellence*, but the objective is the same.

Generally, the members are people who may or may not share a reporting structure or even perform the same role within the organization.

However, each of the members is committed to advancing the skill set within the organization.

The community or network is a great way to engage stakeholders that are interested in actively supporting transformational efforts that directly leverage or affect the skills that they are committed to. It is also a great way to disseminate information and to build a team of champions.

The community contains a leader who will mentor, coach, and direct the members to achieve the desired standard. It also contains members who volunteer for active roles within the network to ensure its success. These members perform the lion's share of the work for determining the standard, providing ongoing training, and mentoring each other.

MENTORS, CHAMPIONS, AND EVANGELISTS

Mentors, champions, and evangelists are people who will take an active role in the transformational initiative. This can either be a positive or a negative role, and these people can be designated by the project team or self-appointed. Either way, their impact on change and adoption must not be underestimated.

Mentors

Mentors are people who are usually resources that report either directly or indirectly into the project. They are often also trainers who are responsible with supporting and coaching people during defined training and one-on-one sessions.

Champions

Champions can be resources that report into the project, but more often than not, they are self-appointed people who will either be pro-change or anti-change based on their own personal level of buy-in. More importantly, these people will have influence over how others perceive the changes and buy into them.

Tip: Finding the most vocal opponent to the changes and *converting* them results in a great champion who will tell everyone how great the changes are going to be.

Evangelists

Just as with champions, evangelists can also be resources that report directly into the project. However, evangelists are typically exclusive promoters of specific aspects of the change—such as a single software package or skill set. They are experts in the area that they represent and they work to ensure that people are convinced to make the changes necessitated by the transformation effort.

WORKSHOPS

Workshops are hands-on learning sessions that can help to provide much needed training of new skills that result from the changes being implemented. Remember that one of the big reasons people resist change is fear. Workshops help to break down that fear by showing people not only the theory behind the changes or new processes, but also how they will participate in executing and managing them on a daily basis.

Training, in and of itself, is a mediocre vehicle for teaching new skills, but when you add in the fear of change and the resistance that people will feel due to their personality types and learning preferences, training on its own is a waste of time and money. Worse yet, it can end up doing more harm than good because it can actually further disengage people by making them feel *talked at* instead of involved.

Okay, that certainly did not sound like the promotion it could have been. Remember that workshops and training are not necessarily one and the same, especially when you consider that training is usually a classroom setting where people read, discuss, and are tested on theory. By contrast, workshops are usually experience-based learning sessions in environments that facilitate trying out and practicing the new skills.

That being said, sometimes change necessitates both. The best option for the change agent is to work with the trainers and facilitators to develop engaging curriculums that leverage hands-on learning as much as possible.

Facilitator's Role

A facilitator is any person who defines and controls the process of a public event—such as speeches, presentations, learning venues, conferences, and workshops. Within the transformation effort, it is the primary role of the change agent to support the change project and often to act as the facilitator during various stakeholder meetings and events. It is critical for the change agent to heavily influence the meeting from the concept to the invitation, to the location and seating arrangements—and to help to manage the flow of information and contributions throughout the entire process.

The change agent is really there to support the change and transformation of the organization and must ensure that care is taken by other participants and project team members to encourage and mediate participation and collaboration of the business organization.

FORUMS

Forums are similar to workshops in that they provide a place for people to come together and discuss the new skills, but they also provide a place for people to come together and discuss the changes to learn more about how they are impacted, what they can expect, and how they can participate.

Great forums provide people with a chance to try out or see the models of their new work routine. All of this is designed to educate people and to get them on board with the transformation. A forum is all about creating an experience and getting feedback.

> Companies that are upgrading office spaces often create a workspace that is designed to look like the new working environment so that people can come and try it out and see what it looks like.

These forums are always supported by mentors who guide people through the experience, answer questions, and ask for feedback. Forums are great for creating a personal image of the changes for people so that they are more than ready to adopt them—they are excited about them. They cannot wait for the changes to be implemented!

GAMIFICATION AND OCM

Gartner defines gamification as: "The use of game mechanics and experience design to digitally engage and motivate people to achieve their goals."[4] In order to fully understand the concept, consider this: game mechanics are the elements of games that are fairly common—such as points, leaderboards, badges, coins, stickers, and other rewards; while experience design describes the journey that players will embark on— such as a quest, plot-line, and world.

In the context of OCM, the change road map provides the experience design, while the activities and other tools utilized can provide the game design. Games, incidentally, are an excellent and nonthreatening way for people to learn new skills in a fun environment.

B.A. Jeopardy

Many years ago, a company decided that its business analysts needed a way to learn new skills in order to ensure that everyone working in this role had the same core understanding of what business analysis was. The new manager decided to host an evening activity which pitted two teams against one another in a game based on the popular television show Jeopardy.

Categories included *Requirements*, *Change Management*, *Process Design*, *Geek History*, and *Geek Mythology*. Each category contained questions that helped to improve the participants' understanding of common terms and definitions. The winning team was awarded coffee mugs and T-shirts, as well as bragging rights. They had so much fun; they forgot that they were afraid of being seen as incompetent if they did not know something. This one activity boosted not only knowledge, but camaraderie and confidence as well.

"Gamification is not always about making games; it's about changing behaviors, engaging people, creating habits, and solving problems in a gameful way."[5]

Gamification is important within the context of organizational change management because the techniques are specifically designed to take advantage of people's innate desires for things like socializing, knowledge, leadership, achievement, status, and individuality. Each of these desires can be aligned to the three needs identified by self-determination theory—relatedness (socializing), competence (knowledge, leadership, achievement, and status), and autonomy (individuality). It is because of this that these techniques have a profound impact on organizational change efforts and transformational initiatives.

Gamification has a wide variety of applications within organizational change simply because of the effect it has on transforming people's levels of engagement, interactions, and participation. Where many texts cite the application of these principles into marketing campaigns, inspirational activities, healthcare, employee productivity, and educational efforts, organizational change management can leverage gamification techniques to improve buy-in, increase participation, transform attitudes, and shift behaviors.

In and of itself, gamification includes the creation of reward and point systems that are made visible to the people who frequently visit websites and perform tasks within specific change management tools. Some examples include:

- Profile completeness score
- Average profile completeness scores of others with similar profiles

- Number of profile visits
- Average number of visits for other people with similar profiles
- Individual time to complete help desk tickets
- Average time to complete help desk tickets of colleagues within the group
- Points for completing change tickets (or other activities) within the time frame
- Average numbers of points held by others within the same work group

How Do You Incorporate Gamification into Your OCM?

In order to begin incorporating gamification into OCM efforts and transformational initiatives, the change agent must start by thinking outside of the box (see Figure 9.2). It is easy to count the numbers of times that a particular website is visited or a survey is taken. However, these two things do not equate to buy-in and will not give you an accurate picture of who's engaged; and more importantly, it will not help you to shift people's attitudes and change their behaviors.

To leverage gamification within organizational change, there must be an element of individuality to the statistics collected and displayed, and there must be a way for the person to influence those statistics. More than that, there must be a way for them to see how they fare against others

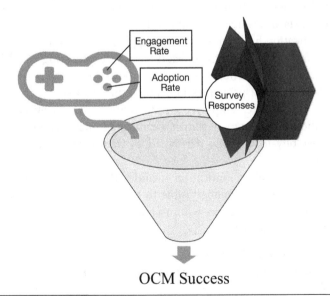

Figure 9.2   Change gamification measures

within the same group. In effect, the change agent is utilizing gamification to reinforce new beliefs, change behaviors, and increase participation by fostering competition.

Luckily, in today's world, many of the common tools utilized for business already contain some type of built-in gamification options. However, change management tools are still few and far between so the question becomes, how do you leverage gamification when you do not have systems with the functionality already built in? Table 9.6 illustrates some examples of the types of data and metrics that the change agent could collect in order to apply gamification techniques throughout the change process, though it is by no means an exhaustive list.

Table 9.6 Gamification

Area	Gamification Measure
Training	• Numbers of attendees • Average training/testing scores of attendees • Individual training/testing scores • Average volumes of training activity participation • Individual volumes of training activity participation
Beliefs and attitudes	• Rewards such as points for doing specific activities • Illustrating volumes of people holding certain attitudes and beliefs about the change (or aspects thereof) (i.e., how many people believe 'x' or want to change?)
Participation	• Individual participation rates across all activities (i.e., how many events, meetings, and activities has this person attended?) • Average participation rates across all activities (i.e., how many events, meetings, and activities have others attended?) • Individual usage patterns for new systems/processes (i.e., is this person using the new system/process?) • Average usage patterns for new systems/processes (i.e., how many people are using the new system/process?)

REFERENCES

1. http://robdkelly.com/blog/leadership/vision-and-mission-state ment-examples/.
2. Garcia, M. L. and Bray, O. H. (1997). "Fundamentals of Technology Roadmapping." Strategic Business Development Department, Sandia National Laboratories.
3. Maurer, Robert, PhD; *One Small Step Can Change Your Life: The Kaizen Way*, Workman Publishing Company. 2014.

4. Burke, Biran; *Gamify: How Gamification Motivates People to Do Extraordinary Things*, Taylor and Francis; 2014.

5 Herger, Mario; *Gamification in Human Resources*, CreateSpace Publishing, 2014.

10

How to Measure Success

There are two types of measures that every change agent must be concerned with: those that assess the situation and enable decisions to guide the transformation; and those that measure the progress of the transformation itself. For the sake of discussion, let's call the ones that assess and enable, *standing measures*; and the ones that measure progress, let's call *shifting measures*.

Standing Measures

Standing measures are those metrics which are fixed and will not change throughout the process of the transformational effort. They are just as important as shifting measures because they help the change agent and the organization make critical decisions up front—during the planning stages of the project—in order to guide the scope, need, and cost of the change management effort.

Shifting Measures

On the other hand, shifting measures are those metrics that will change as the project evolves and progresses from initiation to implementation and beyond. These measures are important because they answer the key alignment question: how much has been done so far? It is the quintessential, "Are we there yet?" inquiry.

Obviously, one of the main differences between these two types of measures goes beyond just the drivers of each. They are also differentiated when they are collected. Since standing measures do not change and will help enable key decisions about the change effort as a whole, they are measured at the outset of the project. And, since shifting measures are intended to track and monitor the progress of project execution, they are measured before, during, and even after the project has concluded.

BEFORE: SETTING UP FOR TRANSFORMATION SUCCESS

Defining Standing Measures

Again, standing measures are those metrics which support the change planning process and enable the change agent to make key decisions throughout the project. In order to define these measures, the change agent should answer the following questions:

- What is the size of the transformation?
- What is the urgency to resolve the problem?
- Who is impacted?
- What are their needs?

Stakeholder Group Size

The more stakeholders in a group, the more diligent the change agent must be in narrowing down the objectives for the overall transformational effort. Remember that each stakeholder will come in with a differing agenda and perspective, and it is the responsibility of the change agent to get consensus among everyone and not manage to the exception.

Need Types and Volumes

Stakeholder need types, as well as the number of stakeholders with specific needs, are important because they will influence the direction and outcome of the initiative. Having this information also helps to identify attitudes that are for and against the changes, as well as helping to understand how those attitudes will evolve over the course of the transformational effort. Therefore, it is important to capture how many stakeholders have specific needs.

Determining Shifting Measures

Let's face it, the ability to measure success in any given initiative is critical to proving that success actually happened. Unfortunately, organizational change management to this point has been missing out on some really important measurements in order to do just that. This, of course, does not mean that these transformational initiatives have not been successful, it is simply that it cannot really be fully proven without looking at a more comprehensive picture.

In the context of organizational change, people often mistake employee happiness with success. However, the truth is that there is really only a loose correlation between employee happiness and the success of a given change effort. It is much like saying there is a rainbow in the sky simply because a person spots a yellow arch.

So how and when do change agents measure success? Let's start with the *when*, before diving into the *how*. The only way to definitively prove change is to measure the differences that occur in predetermined key performance indicators (KPIs) before, during, and after the change efforts have occurred.

Road Trip

Let's say you were going to drive to Los Angeles, California for an important business meeting with a famous movie producer. In order to be successful at this meeting, you need to be well-rested. You pack up your car, but do you start driving? No, not yet, anyway.

First, you have to figure out the best route from your house to LA. Then you have to figure out when your meeting is and how long you have to drive every day in order to get there the day before your meeting so that you get some sleep (which you are really going to need by then) and not blow this once-in-a-lifetime opportunity.

Great! The meeting is in five days, and if you're in San Diego, CA—no problem; but if you're in Cleveland, Ohio, you can count on four, 10- to 12-hour days of driving. (You'd better bring a friend.)

Leveraging the Balanced Scorecard

For the purposes of measuring change, the framework provided by Kaplan and Norton's balanced scorecard provides a really good starting point. "The balanced scorecard is a strategy performance management tool … that can be used by managers to keep track of the execution of activities by the staff within their control and to monitor the consequences arising from these actions."[1]

In other words, the balanced scorecard sets the targets for the overall change initiative. Table 10.1 provides an example of a balanced scorecard for a change initiative.

The balanced scorecard organizes aspects of the change priorities into four categories: financial, customer, internal, and learning and growth. Each of these categories is designed to help the change agent

Table 10.1 Balanced scorecard

Process: Theme:	Objectives	Balanced Scorecard		Action Plan	
		Measure	**Target**	**Initiative**	**Budget**
Financial Perspective: *"If we succeed, how will we look to our shareholders?"*	•	•	•	•	•
Customer Perspective: *"To achieve our vision, how must we appear to our customers?"*	•	•	•	•	•
Internal Perspective: *"To satisfy our customers, with which processes must we excel?"*	•	•	•	•	•
Learning and Growth Perspective: *"To achieve our vision, how must our organization learn and improve?"*	•	•	•	•	•
Strategic Jobs: • Sales • Appt Setter • Acct Manager	•			Total Budget	•

consider what the organization is really trying to achieve from a holistic perspective.

It is important to note here that financial objectives are long term. It can be very difficult to see immediate improvements in the financial position of a company, however, the other categories can serve as indicators of progress toward an improved financial position when you consider that the objectives and measures in these other categories are a part of the cumulative whole of the organization.

This is one of the most likely reasons that financial indicators are not considered and measured as a part of a transformational effort. Either because the change agents believe that someone else is responsible (and doing it), or it is long term and extends well beyond the life of the individual project initiatives.

In each of these categories, the change agent identifies what is changing. In the sample scorecard in Table 10.2, the change agent has cited

Table 10.2 Sample balanced scorecard

Process: Sales

Objectives	Balanced Scorecard		Action Plan	
	Measures	Targets	Initiatives	Budget
Theme: Create long-term sustainable income				
Financial Perspective: *"If we succeed, how will we look to our shareholders?"* • Profitability • Grow number of customers • Best place to work	• Cost of sale • Customer revenue • Free cash • Project costs • Employee satisfaction	• <10% • Sales: $1 million • 1 yr expenses • <50% • >95%		
Customer Perspective: *"To achieve our vision, how must we appear to our customers?"* • Attract and retain customers • Segment market • On-time project delivery	• Number of customers • Number of repeat customers • Number of on-time deliveries	• 4 to 6 • 80% • 98%	• Implement CRM • Customer loyalty program	• $1,000 • $5,000
Internal Perspective: *"To satisfy our customers, with which processes must we excel?"* • Seamless client on-boarding	• Time to on-board	• 2 weeks • <5% variance		
Learning and Growth Perspective: *"To achieve our vision, how must our organization learn and improve?"* • Develop sales skills	• Job readiness	• Yr 1: 75% • Yr 3: 90% • Yr 5: 100%	• Training	• $10,000
• Develop service offerings • Develop KPIs • Develop client support • Align sales and marketing to strategy	• Service ramp-up time	• 2 weeks	• Training	
Strategic Jobs: • Sales • Appt Setter • Acct Manager	• System availability • Strategic awareness • Percent of employee shareholders	• 100% • 100% • 100%	• Profit sharing plan • Marketing system	• 5% total profit • $35,000
			Total Budget	**$51,000**

a *strong financial position*; however, it is recommended to be far more specific and detailed by defining exactly what a strong financial position means—such as *increase overall profitability* or *increase market share.*

The balanced scorecard then captures specific details from each of those categories across four sections: objectives, measures, targets, and initiatives.

The *objectives* identify specifically what will be changed throughout the process of the transformational effort. Again, looking at the example in Table 10.2, the change agent has cited that this transformational effort is being undertaken to change the organization's market share size, amount of revenue, profitability, the cost of customer acquisition, and the total cost of ownership.

Next, the *measures* section identifies how each of the objectives will be measured. For example, if an organization wants to measure changes in revenue, it would use various measures including overall revenue, as well as the revenue for individual product lines or service offerings.

The *targets* section is where the change agent identifies the specific amount—either in dollar value, percentage, or ratio—that the organization wants to change. For example, if an organization wanted to decrease the total cost of operations on its home renovations service offering, the target might be to decrease it by 20%.

Finally, the *initiatives* section of the scorecard is where the change agent identifies the specific project or activities that will be leveraged to achieve the target. In the case of the financial objectives and targets, remember that the evidence of successful outcomes is over a much longer term, so ongoing financial analysis is a perfectly acceptable response. However, the other categories will have short-term activities that can be condensed into a single project more than once.

For example, in order to achieve a target of 99% customer satisfaction rating, the change team has determined that it will employ a combination of initiatives including surveys, mystery shoppers, and a usage analysis to determine how and how often customers are utilizing the new products. These are short term because the KPI can be repeatedly captured in snapshots at various intervals throughout the transformational initiative. These snapshots serve to provide the change management team with a picture of how well the changes are being implemented.

Understanding Baselines and Milestones

Baselines and milestones are key sets of metrics that help to define the starting point (hence the term *baseline*) as well as the specific points where progress will be measured across the transformational effort (*milestones*). Milestones periodically intersect shifting measures at key intervals on the timeline of the project in order to reevaluate the progress. It is the cornerstone of success to have each of these identified while the transformation effort is in the planning stages.

Again, baselines provide critical information about where the organization was when the transformational effort started, and where it is today. This is especially crucial when the organization presents a burning platform to employees and customers as the driving force behind the changes being made, but it is also necessary when it comes to communicating the vision, demonstrating business value, and outlining the goals, objectives, and drivers of change.

Once the baseline has been established, it is relatively simple to reassess some of the key areas and compare them against that baseline in order to determine the progress that has been made. However, before establishing a single baseline or milestone, it is necessary to identify the stakeholders. This is because they will have an influence on the objectives, what things are measured, the targets, and even the initiatives to achieve them.

Impacts of the Problem

It is important to remember that at the outset of the transformational effort, the scorecard must capture the impacts of the problem being faced by the business. These impacts can almost always be summed up into one of the four categories of the balanced scorecard.

Financial

Fiscal business problems are those that have an adverse effect on either the sales or the expenses. In other words, how the problem affects the top and bottom lines. Some of the most common factors that organizations utilize to assess the financial position are:

- Market share
- Revenue
- Profitability
- Customer acquisition
- Total cost of ownership/operations

Customer

Customer problems are those that have an impact on how customers view, relate to, and interact with the organization. This means these problems can very easily detract from the brand and cause customers to spend their money with the competition. Some of the common indicators of problems with customers are:

- Satisfaction
- Loyalty
- Service quality

Internal

Internal problems are those issues surrounding how the organization operates. That is, its processes, policies, and tools utilized to run the business in order to deliver its products and services. Some of the key indicators that there are issues within an organization include:

- Culture
- Product or service quality

Learning and Growth

Finally, the learning and growth category is entirely focused on employees. Problems in these two critical areas will indicate serious issues throughout the organization and will most certainly accompany poor financial performance and low customer satisfaction scores. These indicators include:

- Emotional well-being of the employees
- Capability of the employees to deliver specific products or services
- Capacity of those products and services which can be delivered within a specific time frame

CHANGE SCORECARD

The change scorecard is a method for predicting the ability of an organization to change. As such, it contains measures that are both standing and shifting. This is so that it can be leveraged to reevaluate the organization throughout the transformation and after it is completed.

As discussed in Chapter 7, the change scorecard seeks to answer four measurable questions. These questions are depicted in Table 10.3.

In answering these questions, this scorecard effectively provides a very detailed look at the changes within the context of the business and its people, processes, and technology (as illustrated in Table 10.4). The results of this detailed scorecard are rolled up into the overall change score depicted in Table 10.5.

Table 10.3   Change drivers

Point	Measurement Type	Purpose
What is changing, by how much, and over what time period?	Shifting measure	Helps the change agent set a baseline and track progress against that baseline.
How difficult is that change going to be?	Standing measure	Helps the change agent determine how long transformation is likely to take and could identify the means to mitigate issues and reduce that time frame.
How critical is it for the organization to make the changes?	Standing measure	Helps the change agent determine the types of techniques that should be used to propel change at the appropriate pace.
How ready is the organization to make the necessary changes?	Shifting measure	Helps the change agent determine the best approach and techniques that should be used to propel change at the appropriate pace.

Table 10.4   Change scorecard

Item	Factor	Definition
Degree of change	People	Who is impacted by the changes?
	Process	What processes and methods will change the way in which the people impacted perform their jobs?
	Technology	What tools and technologies will change, and will they change how people perform their work?
Complexity	Functional complexity	How complex is the new technology that is being introduced?
	Audit complexity	How critical is it to audit the work performed by this technology?
Criticality	Business criticality	How important is it to the business that this technology be implemented?
Readiness	Barrier complexity	What kinds of barriers exist for this change?
	Barrier criticality	How difficult will it be to overcome each barrier?
	Barrier mitigation strategy	Is a comprehensive strategy in place to help overcome the barriers?
	Mitigation complexity	How complex is it to implement the mitigation strategy?
	Alternative solutions	Have we identified different solutions to help mitigate the barriers identified?

Table 10.5 Change score (the Likert scale is used for scaling responses in research)

Factor	Description	Score	Likert	Individual Weighted Score 100%	Maximum Weighted Item Score 100%	Final Weighted Change Score 100%
Degree of change	Percent of change that must occur	68%	3	50%	20%	10.00
Complexity	How difficult the changes will be	High	5	100%	25%	25.00
Criticality	How important this change is to the business	Important	3	50%	30%	15.00
Business readiness	How prepared the business is to tackle them	95	5	100%	25%	25.00
						75.00

Key Activities

The activities in developing the change scorecard include understanding how to describe the responses to the four questions in terms of tangible KPIs.

Degree of Change (What Is Changing, by How Much, and over What Time Period?)

To determine what is changing, by how much, and over what time period, some simple math is involved. Let's break the question into those three components (see Table 10.6):

- What is impacted: represents the total amount of people, processes, and technology that will be impacted by the transformation initiative.
- By how much (degree of change): is a simple count of the total number of people, processes, and technology impacted versus the total number in the organization. This represents the overall size of the change or transformation within the context of the organization as a whole.
- What time period: indicates the overall time period for the transformation to occur, described in months.

Each of these columns (except the degree of change) is tallied in order to establish the overall scale of the transformation. The degree of change is calculated by using the totals from the second column and dividing it by the first column and converting the result to a percentage.

Example

A hydroelectric company is changing its multiple operating systems in favor of a single platform. The entire organization has 5,300 people (employees and contractors), 6,300 computer systems, and 300,000 processes listed in its process library.

Of this, all 5,300 people are expected to switch to the new operating system, on 5,300 computers. This means that approximately 4,500 processes will change as a result. However, only 345 are expected to be changed by the implementation project, and the project will take one year to complete. Table 10.7 illustrates this example.

Functional Complexity (How Difficult Is that Change Going to Be?)

The functional complexity is a highly valuable tool for assessing the complexity of the changes being implemented. As shown in Table 10.8,

Table 10.6   Degree of change matrix

	Total Number	Number Impacted	Degree of Change	Time Period	Percent Changed to Date
People	Total number of all people in the organization, customer base, and vendor talent pool.	Estimated total of those impacted by the changes. How many of the total will be asked to change?	Overall target percentage of people that will change.	Describe the time period in months.	An ongoing KPI for measuring progress against the number impacted.
Process	Total number of processes utilized across the organization or group.	Total number of processes that will be changed or introduced.	Overall target percentage of processes that will change.	Describe the time period in months.	Rate of change measured as a KPI.
Technology	Total number of systems utilized by the organization or group.	Total number of systems that will be changed or introduced.	Overall target percentage of systems that will change.	Describe the time period in months.	How much change has occurred to date?
Total	Total of this column.	Total of this column.	Total percent of all of the changes.	Total amount of time or the greatest number if happening concurrently.	Total of this column.

Table 10.7 Sample degree of change matrix

	Total Number	Number Impacted	Degree of Change	Time Period	Percent Changed to Date
People	5,300	5,300	100%	12 months	0 / 0
Process	4,500	345	7%	12 months	0 / 0
Technology	6,300	5,300	84%	11 months	0 / 0
Total	16,100	10,945	68%	12 months	0 / 0

it contains an assessment of the overall changes by comparing values for each attribute of the new system, process, or program. These include:

- Business criticality.
- Number of integration points with other systems, processes, or programs.
- The overall number of anticipated users.
- The number of processes incorporated into the new system or program.
- The volume of transactions that the new system is expected to process.
- The amount of financial calculations that the new system will perform.
- The amount of other calculations that the system will perform.
- How much does this new system, process, or program have an impact on safety?
- How much does this new system, process, or program have an impact on security?
- What is the audit complexity?
- What types of applications will interact with and interface with this new system, process, or program?
- What types of transactions will the new system, process, or program manage?

How Is It Conducted?

The change agent, in collaboration with the rest of the project leadership, determines the best response for each of the answers compiled from the previous list on a scale from low to high. Using the functional complexity estimation tool, again referring back to Table 10.8, the tool then completes the calculation of the complexity for the changes.

Table 10.8 Functional complexity matrix

Simple-Low	Low 1	L-M 2	Moderate 3	M-H 4	High 5	Weighted Complexity Score 100%	Maximum Weighted Item Score 100%	Final Weighted Complexity Score 100%
Business criticality						0	10%	0.00
Number of integration points						0	2.50%	0.00
Number of users						0	2.50%	0.00
Number of processes						0	5%	0.00
Number of transactions						0	5%	0.00
Number of financial calculations						0	15%	0.00
Number of calculations						0	10%	0.00
Safety						0	10%	0.00
Security						0	10%	0.00
Audit						0	10%	0.00
Integrated App Types						0	10%	0.00
Transaction Types						0	10%	0.00
Response Analysis						0		
Complexity Rating					High			0

It is important to note here that this tool also provides the functional complexity for the project requirements as there is a direct correlation between the complexity of the system, process, or program being implemented and the complexity of the changes being made. The more complex the result, the more effort that will be required to support people through the changes being implemented.

Audit Complexity

The audit complexity is one of the primary subcomponents of the functional complexity assessment of the new system, process, or program. Together with business criticality, it helps to determine some of the critical considerations in the development and implementation of the overall solution. It does this by ensuring that those implementing the solution think about all of the factors of the solution that would make it more or less difficult to design, build, and deploy.

No one really likes the idea of an audit, or building some audit capability into a new solution, because it makes that solution more time-consuming to use. However, there are systems and programs that require some kind of traceability of the transactions and processing being performed. The audit complexity assessment is designed to understand the level to which this functionality must be implemented in order to have a full end-to-end solution.

On a scale of one to five, the assessment considers all aspects of the overall program, system, or processes, and rates it according to the criteria shown in Table 10.9.

Table 10.9 Audit complexity

Valid Codes	Criteria
1—External	Mandatory external audits and traceability required for regulatory purposes.
2—Analytics	Validates calculations and outputs, generates analytics using the outputs.
3—Internal controls	Validates calculations and outputs, generates reports and controls the process.
4—Reports	Validates calculations and outputs, generates management reports.
5—System	Simple calculation and outputs validation to verify and control the process and reduce errors.

Business Criticality (How Critical Is It for the Organization to Make the Changes?)

Business criticality, on the other hand, is an assessment of how important the program, system, or processes are to the business. Business criticality is often some vague and intangible idea based on individual inputs from disparate stakeholders. However, by conducting this assessment, the change agent, together with the project team, is able to set a baseline for what the project results really mean to the business as a whole.

On a scale of one to five, the assessment considers all aspects of the overall program, system, or processes, and rates it according to the criteria shown in Table 10.10.

Business Readiness (How Ready Is the Business to Make This Change?)

The ultimate question for every change agent is, "How ready is the business for this change to be implemented?" Change management is not only about how well and how quickly people adapt, but how willing they are to do it. However, willingness is a by-product of many things including overall organizational preparedness. The more the people impacted believe that the organization is not prepared for change, the less motivated they are to adapt.

In order to fully understand business readiness, it is therefore crucial to measure both willingness and preparedness. The question is how? Preparedness is the easy part because preparedness is tangible. Willingness, on the other hand, can get a bit tricky. Willingness is often seen as an intangible because it is a *feeling* that people have—and that feeling can be influenced by factors outside of the care and control of the change agent.

Table 10.10   Business criticality

Valid Codes	Criteria
1—Mandated	Will result in regulatory noncompliance and potential legal or financial penalties if not implemented.
2—Critical	Product will not meet customer needs if not implemented.
3—Important	May adversely affect customer or user satisfaction if not implemented.
4—Useful	No significant customer or functional impact if not implemented.
5—Wish	No impact but will increase customer perception of value if implemented.

Preparedness

Organizational preparedness is all about the systems and processes that must be in place before the transformation can even begin. Realistically, this could mean that these items are prepared after the project has begun, but before the applicable work can be started or completed. For example, before a new software application can be installed onto a company's computer systems, the software must be purchased, then tested on the company's standard computers. Those computers must have the right amount of space available and the right operating system installed. This is measured by identifying and simply counting the items that must be in place before the transformation can start. This is the baseline for preparedness.

Willingness

Willingness is a measure of how cooperative people will be during the transformation. For example, a group of friends would like to go see a movie on Friday, but before they can go they must all decide on which movie, and not everyone is interested in seeing the same movie. So, they are spending a lot of time negotiating which particular movie they would all enjoy. This is measured by conducting surveys, tracking engagement rates, and tracking adoption rates. This is the baseline for willingness.

DURING: REEVALUATING PROGRESS AND ACHIEVING MILESTONES

Throughout the process of transformation, the change agent will work to track and monitor changes in attitudes and compliance to the changes, and then compare that against the baselines as well as the anticipated outcome.

Degree of Change

The degree of change is the measure of what is changing, by how much, and over what time period. This can be reevaluated at each milestone to track progress. However, realistically, it is better measured on a routine basis such as weekly or biweekly in order to closely monitor that progress and identify trends and problem areas. This will enable the change agent to maintain alignment with the established plans and ensure a greater chance of success.

Business Readiness

Preparedness

Again, preparedness measures the quantity of systems and processes that must be in place before the transformation can begin. Throughout the process of change this can be reevaluated to determine if any of the items are completed or no longer required. This can be measured at key milestones and should be reported on as long as they are outstanding. This will enable the change agent to ensure that systems and processes are appropriately implemented.

Willingness

Willingness measures the degree to which people intend to cooperate with the transformation. Throughout the process of transformation, this can be reevaluated to determine trends and to identify pockets of resistance. Just like the degree of change, this can be reevaluated at each milestone to track progress. It is best evaluated routinely so that progress can be closely monitored and problem areas identified. This enables the change agent to better support people throughout the changes. As previously stated, willingness is measured by tracking adoption and engagement rates, as well as conducting surveys.

Adoption Rate

The adoption rate is the number of people who are applying the new solution (processes and/or tools) within a given period of time. For example, as illustrated in Table 10.11, if 50 people are using the new software, but 100 software applications have been installed in five months, 50% of the intended participants have adopted the changes. The adoption rate is 10 people per month.

The formula for calculating the adoption rate is:

Number of people using the solution/Number
of solutions implemented = Result/100
Result/Time elapsed = Adoption rate

Compliance is a factor of adoption since it can be assumed that the number of people using the new solution is in compliance with the organization's new policies and procedures.

Engagement Rate

The engagement rate is the volume of people who regularly attend scheduled meetings and events. Quite simply, it is a count of the number of

Table 10.11   Adoption rate

Item	Count	Result	Rate
Time elapsed	5 months		
Number of people actively using the software	50	50%	10 people per month
Number of software applications implemented	100		

people required and invited to project meetings and events, compared to the number of people who actually show up and participate. For example, as shown in Table 10.12, if 100 people are invited to 12 meetings over 6 months, but an average of only 35 people attend, 35% of the people required participated. The engagement rate is 70 people per month.

The formula for calculating the engagement rate is:

$$\text{Number of people attending events/Number of people invited} = \text{Result}/100$$
$$\text{Result/Time elapsed} = \text{Engagement rate}$$

Attitudes

Many people believe that the single best method for determining a shift in attitude is to conduct surveys of the people involved. After all, aren't they going to be able to describe how they feel about a particular subject over time? The answer is not a simple *yes* or *no*. The truth is that people complete surveys based on how they are feeling in that exact moment in time, not just in general. That means that other things can impact how they respond to the survey, depending on how they feel when they take it.

Attitude is complex. Wikipedia defines it as "an evaluation of an object with some degree of positivity or negativity."[2] However, in reality, attitude translates that positive or negative feeling directly into action.

This is great news for change agents because it means that they can actually measure actions in order to get a sense of how attitudes are shifting throughout the transformation process. When combined with the results of surveys, it paints a powerful picture of the change efforts.

Table 10.12   Engagement rate

Item	Count	Result	Rate
Time elapsed	6 months		
Number of people attending events	35	35%	70 people per month
Number of people invited	100		

For example, when the results of a survey show that people are less than pleased with a change being implemented, and that result is combined with how quickly people respond to meeting requests, complete tasks assigned to them, or even to enroll in the training, it becomes clear that people's attitudes are negative. However, when people begin to respond faster and complete their assigned tasks more readily, and that is coupled with a survey that illustrates a middle of the road feeling about the changes, it becomes obvious that progress is being made in shifting people's attitudes in a positive direction.

AFTER: VERIFY AND FINALIZE RESULTS

In closing the transformation project, the change agent must report on the overall effectiveness of the change management tasks to the sponsors. In doing so, the change agents demonstrate the value by providing clear costs for specific returns.

Change management as a separate set of tasks within the overall project are often a hard sell to sponsors who want to get tangible deliverables for their investment. Change agents must be able to demonstrate value to the sponsors and business stakeholders in order to help them understand the true impacts of organizational change management. In order to accomplish this, the change agent must be able to answer the following questions:

- What was accomplished?
- What techniques were the most successful?
- What techniques were the least successful?
- How much did the change management activities cost?

What Was Accomplished?

In order to determine what was accomplished, the change agent reevaluates the overall progress made to date against all of the baselines that were gathered and measured throughout the project. These include both short- and long-term metrics. Again, these are calculating changes as follows:

- Differences in product or service quality (+/−)
- Differences in customer satisfaction scores (+/−)
- Differences in market share (+/−)
- Differences in revenue (+/−)
- Differences in profitability (+/−)

- Differences in the cost of customer acquisition (+/−)
- Differences in the total cost of ownership (+/−)
- The degree of change

Which Techniques Were the Most/Least Successful?

It goes without saying that not every meeting will attract every invitee, and not every event will be heralded by participants as *the best workshop ever!* To understand which techniques were the most or least successful, the change agent calculates how many people attended each of the meetings or events, and then determines if the rates of adoption increased or decreased in the time period immediately following each event.

- Calculate the engagement rate of individual events
- Compare the ongoing adoption rate track record against the event timeline
- Conduct closing surveys

How Much Did the Change Management Activities Cost?

Calculating the overall cost of the change management activities is fairly simple. The problem is that most sponsors, project managers, and change agents do not calculate it. Cost is a factor of time and the number of resources utilized.

The most effective way to demonstrate value is to calculate the cost of individual activities and events, and then tabulate the sum total. This can be further used to calculate the return on investment (ROI) of overall change management. Of course, calculating this will require support from the sponsors.

The formula for calculating the ROI of organizational change management activities is:

$$ROI = (\text{Gain from the investment} - \text{Cost of the investment})/\text{Cost of the investment}$$

In this case, the gains from the investment are those activities and events that directly correlated to an increase in adoption. Finally, the change agent can work with the sponsor to determine specifically where that gain occurred.

- Increased product or service quality
- Increased customer satisfaction scores

- Increased market share
- Increased revenue
- Increased profitability
- Decreased customer acquisition costs

REFERENCES

1. Kaplan, Norton; *"2GC Balanced Scorecard Usage Survey." 2GC Active Management.* Retrieved 28 May 2014.
2. https://en.wikipedia.org/wiki/Attitude.

11

Creating a Winning Organizational Change Management Strategy

A strategy is an action plan that is designed to achieve a major or overall goal or objective. This is unlike a methodology that is a system or collection of methods utilized in a particular field to achieve a specific goal. In the field of change management, the strategy describes the major driving factors behind the transformation, as well as the overarching plan for its completion.

Both a strategy and a methodology are recommended for change management efforts, particularly on large-scale transformations because each will help to focus the team and the organization in differing and complementary ways. The change strategy will help to answer the following three basic questions:

- Where is the organization right now?
- Where do the executive stakeholders want or need to take the organization?
- What is needed from this specific transformation initiative to get there?

Where is the organization right now? To answer this question, it is important to understand as much about the organization as possible—including how it operates internally, what drives its profitability, and how it compares with competitors. This is precisely where and why the information about the organizational climate and ecosystem comes in.

Where do the executive stakeholders want or need to take the organization? In answering this next question, the change agent must work with the key stakeholders to set out the top-level objectives for the initiative. The change agent supports and guides those stakeholders to develop the transformational vision, mission, objectives, values, and goals. Where do they see the business at the end of this initiative?

What is needed from this specific transformation initiative to do to get there? Finally, the change agent works with the project team to determine what changes need to be made in order to deliver on those strategic objectives, and to determine the best way of implementing those changes. They also work with the stakeholders to outline the changes to the structure and the financing of the change management that will be required and to determine the goals and deadlines you will need to set for yourself and others in the business.

The change strategy is important because it answers these three basic questions at the outset of the planning process and derives the plan of action from there. To this end, the strategic plan is really the culmination of key information that sets the parameters and determines why key methodology decisions have been made. This key information includes knowledge of the business climate and ecosystem, the level of business readiness, and the budget.

It is all too common that change management is an afterthought, and this is reflected in the financial commitment of the organization to the project. While it can be said that this is really a part of a much larger issue in skewed organizational values, it is important to note that an organization that cannot assign a real functioning budget to their change efforts is doomed to fail and will consistently obtain less than desirable results.

When change is an afterthought, it is also present in the business climate in that it shows a lack of respect the organization has for its employees, customers, vendors, and partners. People in the organization know and feel whether they are valued and this is demonstrated through the climate and the culture that immerses them. Based on this, it's fair to say that the business climate, coupled with the ecosystem and business readiness, determines the appropriate methodology selection and ultimately results in transformational success.

Formula for Change Success

Business climate + ecosystem + business readiness determines methodology and results in transformational success (see Figure 11.1).

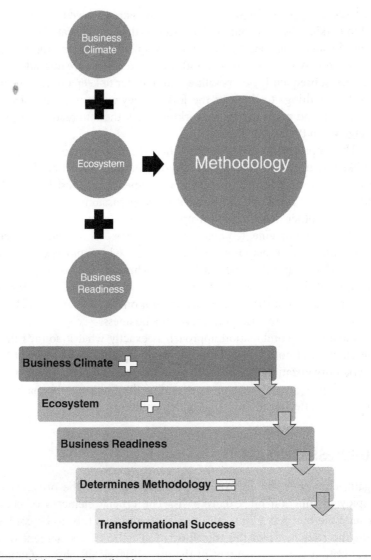

Figures 11.1   Transformational success formula

Change strategy can best be utilized via one of three approaches, and the work done to develop that strategy should support that recommended outcome. These approaches are:

- Top-down
- Bottom-up
- Combination

A top-down approach is when the executives are involved in the change and in making the decisions; and are seen as driving the transformational efforts by adopting changes first. For example, if upgrading the versions of Microsoft Word from one to another, the executives are upgraded first, and then subsequently others follow suit in order to keep up with the *boss* and deliver things to them in the format they are using. This strongly encourages and motivates those working under them to readily adopt the change when their turn comes.

This approach is very effective when there is resistance and people are likely to not be reasonable throughout the transformation. It is often seen as a directive from the upper management that must be complied with under the threat of job loss or other repercussions.

On the other hand, a bottom-up approach is when the rank-and-file employees are implemented before the executives. This is not to be confused with a grassroots movement in which those employees spawn the idea for change and then drive it forward. This approach is very effective when people have high levels of buy-in and there is a lot of trust from the upper management that employees will not only cooperate with change, but will drive it in the best interests of the business.

Finally, the combination approach is exactly what it sounds like: a combination of both top-down and bottom-up approaches. It is utilized in larger organizations—those with a lot of flux and fluidity or when there are varying degrees of urgency surrounding the changes to be made. This approach is effective at reducing implementation time.

BUSINESS CLIMATE

Again, climate is the environment/culture in which the organization's employees function on a daily basis. The crucial elements of climate include the business vision and mission statements (not to be confused with the statements of the transformation efforts); a statement of the organization type; a statement of the overall levels of job satisfaction; a description of the culture; a statement of the strength of the organization's brand; and a statement on customer loyalty.

How Does It Tie into the Strategy?

As previously stated, the business climate helps to determine the methodology selection. Suppose, for example, that the change agent was building out a strategic plan for an organization that had a lot of infighting, poor process compliance, high attrition and absenteeism, low customer retention rates, and a low brand recognition score. What would happen if that change agent selected a methodology that was heavily participation-oriented and required input from everyone in order to proceed with implementing any decisions? Obviously, the change agent would struggle to no end on this initiative because one of two things would happen. First, people who never get a say will suddenly have one and want to control *everything* they possibly can just to be obstinate. Or, second, no one will bother because they feel as though their input does not really matter anyway.

If change strategy can really be summed up into top-down, bottom-up, or a combination approach, then the knowledge of the business climate must dictate that approach through the selected methodology. In other words, one of the approaches will help to determine the best methodology to achieve the desired results. It begs the question: how much influence does this initiative require from executives in order to fully realize the desired transformation?

ECOSYSTEM

As discussed in Chapter 5, the business ecosystem is the summarization of the operational infrastructure and business model. The elements of the ecosystem are the business model, a summarization of the organization's key relationships, and a description of the organizational structure.

How Does It Tie into the Strategy?

Again, the business ecosystem, climate, and readiness are required to select an appropriate methodology for executing the transformation. Unlike the business climate, which contributes to the selection by describing how the people are likely to react to change based on their attitudes toward each other and toward the organization, knowledge of the ecosystem contributes by describing how the business operates.

This informs the change agents about which methodology is likely to contain steps and stages similar to those already employed. Using similar steps and stages enables the agents to select methods and processes that are less likely to be rejected outright by the organization because they are similar to those already utilized for other functions.

The truth is that the change itself is going to cause a lot of angst; the methodology should not. It should feel familiar so it is easier to follow and trust.

BUSINESS READINESS

Business readiness is the preparedness of the organization to step into and adopt the new systems and/or processes as designed. The core areas of business readiness are people, process, and technology.

How Does It Tie into the Strategy?

Business readiness is critical in determining the methodology utilized because knowing how well prepared the organization is, helps to determine what steps and stages are necessary to accomplish the transformation within the time period required. By understanding what tasks and activities must be accomplished in order to fully prepare the organization for the transformation, the change agent is able to identify steps that must be included in the methodology to support and enhance this preparedness.

In addition to supporting the selection of the appropriate methodology, business readiness contributes to the level of confidence that people will have in the ability to change, as well as in the team that is implementing those changes. It accomplishes this because when people can see that the necessary tasks have been completed before implementation, they will trust that the implementation will go smoothly. Consequently, if they see that important tasks are left undone, they will not have confidence in the implementation, and this will drive down their contribution levels, as well as their morale.

Moving Day

Several years ago, a large transportation company was moving its employees from one large corporate office space into a larger and more modern location. Everyone was looking forward to the new location. On moving day, every employee who was scheduled to relocate packed up their desk contents and went home for the night.

The next morning, they arrived at the new office location and began to unpack. However, as more and more people tried to set up their computers and telephones, they discovered that there were no electrical connections, Ethernet ports, or telephone lines set up in any of the cubicles.

After some chaos and some irate phone calls, everyone repacked their belongings and returned to their original office for another month to allow the lines and connections to be installed.

Needless to say, in the aforementioned situation, the people were very skeptical about the team's readiness when the next moving day arrived, so several groups sent a representative to check each and every cubicle to verify that the items had been installed correctly before they considered packing again. Their trust in the implementation team was lost.

Business readiness does more than just support methodology selection and foster confidence, it also supports the selection of appropriate goals for the transformation. When key factors are not ready for implementation in the anticipated time frame, additional change management goals must be established in order to ensure that people's confidence and trust is maintained and does not diminish while they wait. It is impractical to suggest that in any given scenario every person will be mentally, emotionally, and cognitively ready for change when it happens; and it is unrealistic to believe that technology and process transformations can occur when people are not prepared.

ELEMENTS OF A WINNING ORGANIZATIONAL CHANGE MANAGEMENT STRATEGIC PLAN

The cornerstone of every strategic plan is the means of tying together the vision with the action needed to achieve that vision. An organizational change strategic plan is no different. To this end, the main elements of this plan include the following steps:

- Laying the groundwork
- Outlining transformational goals, objectives, and primary focus areas for the change efforts
- Determining change initiatives to meet the transformational objectives
- Selecting the change methodology
- Designing the communication architecture
- Determining measurement and accountability

Laying the Groundwork

This is the work of discovery and getting to know the organization in order to understand the business context of the changes, defining the vision, identifying the key stakeholders, defining the high-level transformation requirements, and conducting the strengths, weaknesses, opportunities, and threats (SWOT) analysis.

Laying the groundwork for any transformation effort is a crucial part of developing the change strategy because it literally sets the foundation for a successful initiative. Every task in this step is critical to establishing a solid change strategy that will ensure full adoption. Understanding the business context of change provides insight into how transformation can be possible and readily adopted. The vision will ensure that people have the ability to retain focus on the end goal when they get into the weeds. Identifying the stakeholders will make certain that every team is represented throughout the process. The high-level requirements that detail everything that must change will be accounted for and the SWOT analysis will ensure that the transformation initiative is looked at through a critical lens that may provide additional insight and information to support its success.

Key Activities

- Mapping the business context of the changes
- Determining vision, mission, and objectives
- Identifying the key stakeholders
- Defining the high-level transformation requirements
- Conducting the SWOT analysis

Goals, Objectives, and Primary Focus Areas

The strategic goal-setting task is where the overall transformation targets and objectives are set. It is used to define what will be changing during this particular transformation, even if it is a part of a much larger, multi-year effort to transform the organization.

Sample Strategic Goal

An electric utility company decided to migrate from multiple platforms to a single operating system. The sponsors felt, during the strategic planning stage, that not every system could possibly be migrated over to the new operating system so they applied the 80/20 rule. They believed that only 80% of all systems would ultimately be upgraded because of various factors related to the employees, the geographic locations, and the systems themselves.
They made 80% the target migration rate to aim for.

However, these goals are not just about how many people and/or systems or processes will be transformed. It is about determining the right objectives that are realistic given the current situation and the other influencing factors.

Goals could be:

- The number of people/systems/processes that will be changed
- When the people/systems/processes will be changed

Objectives could be:

- Change the way that employees give feedback on organizational performance throughout the change
- Provide employees with the opportunity to test out the new workspace before it is implemented
- Ensure that as many employees feel welcomed to participate and contribute to the changes as possible

Primary areas of focus could be:

- Communication forums
- Buy-in rates

Every successful initiative has some form of goals, objectives, or areas of focus because this enables the people participating, and those expected to buy in, to individually manage the alignment of their activities to the vision and to change of their own volition and at their own pace. There isn't a successful sports team around that does not understand that when they set goals, it is easier to achieve them because players can put themselves in the right places to support the objective. They will do the extra work, move themselves on the playing field, and learn what they need to do to prepare themselves for the next play that might help the team win.

Key Activities

- Strategic goal-setting

Strategic goals are the end results that will measure the overall success of the transformation efforts. These goals are best defined by answering questions that include: what are the key change issues the effort will face and what strategic change goals need to be established?

Determine Change Initiatives

Change initiatives are those events that will support the transformation by educating, breaking old habits, building confidence, and enabling a cultural shift. These events could include workshops, training, team building activities, and even social media campaigns.

Sponsored change initiatives are critical to success because they provide crucial hands-on opportunities that the people impacted by change and those who are expected to carry forward the new systems and processes with knowledge and confidence to be successful. No change effort can be considered whole or complete without initiatives to engage people at all levels of the organization.

Key Activities

- Identify change initiatives

The identification of specific change initiatives is the development of ideas and processes that will lead to the achievement of the strategic goals. This is accomplished by answering the following questions: "What are some other typical change initiatives?" and "How can the project team enhance risk governance and oversight throughout the life cycle?"

- Develop action plans

This step defines what needs to be done to implement the change initiatives. This area of the planning process tends to present the biggest challenges to most organizations when it comes to change. The detail needs to be sufficient to guide the implementation process and yet most plans are prone to being too vague or not well thought out. At a minimum, action plans that are the most effective have the following characteristics:

- They clearly state the change initiative
- All major events, phases, and tasks are associated with key performance indicators (KPIs)
- They establish the person responsible for each action item, as well as the supporting staff
- The scheduled start date and anticipated implementation period are set
- Resources, such as capital, operating funds, and staffing hours are stated
- Feedback expectations and tracking are determined within the plan itself

- Define the change road map: establish priorities

Finally, this step defines how the action plans will be implemented and determines the order of precedence. It is important here to set appropriate and realistic priorities for the road map because not doing so will erode the credibility of the change agent and make it harder to gain buy-in.

Methodology/Approach Selection

The change agent is responsible for selecting an appropriate methodology or approach based on the vision, objectives, risks, issues, business readiness, and the desired change initiatives for the transformation effort.

In addition to the strategic change plan, the methodology defines specific activities and deliverables to manage the process of transformation. It is crucial to have both of these in place, as one does not replace the other. The methodology identifies change-specific deliverables and tasks to ensure that the transformation is holistic and considers all possible paths to buy-in.

Selecting the appropriate methodology is simple when the decision is based on: where the organization currently is; where it wants to go; what must be accomplished along the way and in what order; how ready is it to start the transformational journey; and what tasks, activities, and deliverables are required to ensure that the organization achieves the changes to the highest degree possible? Table 11.1 illustrates a matrix-style decision table to support the selection of the appropriate methodology based on all of the relevant factors.

Key Activities

- Review the organization's current state and compare that to the future state (the target)

In essence, this is similar to conducting a gap analysis that is typically done by business analysts. This step begs the question: where are we and where are we trying to go? Think back to the road trip analogy.

- Review the specific requirements for the change to ensure its success

Here, the change agent consumes and digests the information that was provided by the stakeholders about their needs and preferences, as well as the specific details about the change itself, such as the new processes or systems. This step begs the question: "What do we need to do to get to our destination?" The change agent leverages the information acquired previously to determine any change-specific requirements that they will have to account for to ensure success.

- Review the level of business readiness

By now the change agent should understand how prepared the organization is to make this change. The change agent should be the champion for that preparedness and ensure that the project does not move ahead

Table 11.1 Recommended methodology matrix

Recommended Methodology	Business Context (where the organization is now)		Vision (where the organization wants to go)		Business Readiness (how ready is it to start the transformational journey)		
	Climate (level of support)	Ecosystem (strength)	Vision (clarity)	Change objectives (clarity)	People (level of buy-in)	Process	Technology
ADKAR	Low	High	High	High	High	High	High
Kotter	Moderate	Low	Low	Moderate	Moderate	Moderate	High
IIEMO	Moderate	Moderate	Moderate	High	Moderate	Moderate	Moderate
AIDA	Low	High	High	High	High	High	High

without carefully thinking about what is needed first. When things are pushed ahead to meet a schedule in spite of readiness, the project loses a lot of hard-fought credibility that can be difficult to recover.

- Select the change initiatives

All of the techniques described in Chapter 9 are suggestions for the kinds of things that should be in the change agent's toolbox, ready to be implemented under the right circumstances on the right transformation project. It is important to select those initiatives that will best engage the people impacted by change in ways that are meaningful to them. This will go a long way in reducing the stress of change and making the project outcomes that much easier to adopt.

- Select the best approach

The approaches described in Chapter 10 lay out the various types of guidelines that can be leveraged to successfully implement change. It is the task of the change agent to select the most appropriate approach based on all of the information that they have about the organization, its people, and the desired vision.

- Build the methodology

As previously discussed in Chapter 10, an approach is not a methodology. While to many they may look similar, the methodology requires some kind of governance and validation processes. These can, however, be added to any of the selected approaches to fill this gap and ensure its completeness.

Change Governance: Measurement and Accountability

Governance is the task of managing and controlling the process in order to mitigate risk and deliver consistent and predictable results. Change governance is exactly that; however, it is not to be confused with change control. Where change control is a technical governance to control the changes made to systems from a configuration or a release management perspective, change governance ensures consistency is applied to the techniques of organizational change management in order to predict and control the outcomes of transformation efforts.

Change governance is important not only because it mitigates risk and delivers consistent results, but also because it provides a clear method for managing organizational change initiatives throughout the life cycle of the transformation process. In other words, it helps the change manager better understand how well the team delivers; how well the applied techniques perform; and it works to monitor and document both the successes and shortcomings of the change process.

Key Activities

- Define target values for the parameters

Think of parameters as boundaries. A target value is the goal value for the parameters set for monitoring the process.

Mass Migration

A company was migrating its computer systems to a new operating system. It had determined that, of the 6,000 target systems, they would be satisfied if somewhere between 65% and 74% of these systems ultimately ended up being migrated. They effectively created a target range that set parameters for the project.

- Define KPIs to be measured

KPIs are the metrics that provide a detailed view of how well the processes are being conducted. In this case, the change agent sets the KPIs that will determine if the transformation is on schedule, heading in the right direction, or in desperate need of a makeover. This is not the same as tracking the physical progress—it is ensuring that people are adopting the systems as expected and their mindsets are shifting toward the positive.

- Identify routine intervals (evaluation timeline)

Finally, the change agent is responsible for determining when the metrics will be collected and reported. This schedule will form a crucial part of the strategy and ensure that the project maintains transparency with the stakeholders about hold-out groups or outright change resistance so that it can be managed in a timely manner to ensure the successful transformation of the organization.

12

Adopting a Culture of Change

A culture of change is not the same thing as constant, unpredictable, and unmanaged change. A culture of change is where people readily accept change as a part of the routine. They predict it and expect it. They accept it because change is well managed and they can trust the organization and its leadership.

Remember that culture is effectively the way of life for the people who work within the organization in that it is the shared beliefs, practices, and behaviors that are the social norm. A culture of change, on the other hand, is the overall way in which people accept and adapt to change on a daily basis. It is the way in which the attitude toward change is woven into the fabric of the culture of that organization.

Change that is constant, unpredictable, and unmanaged is chaos. It is subject to the whim of organizational management; and people find it hard to trust when things can change on the fly with little forethought, planning, or consideration for the impacts.

Organizations demand increasing fluidity in order to keep up with competitive forces within the industry and the marketplace. Technology is changing at such a fast pace that by the time the new cutting-edge technology has been implemented, it is already obsolete. It is this fluidity that leaves some people feeling as though they are change-weary. Before discussing how to build it, let's first discuss what a culture of change looks like.

ATTRIBUTES OF A CHANGE-READY CULTURE

Touted as a new organizational utopia, a culture of change is one where the people readily adopt change and willingly embrace it. It's a great concept—and one that executives everywhere are pining for. The trouble is that this culture is a reflection of leadership, not the people who follow it. Now, this ideal is not necessarily a bad thing to aspire to. In fact, it's a pretty great ideal when one thinks about what it really means and why it is important to build.

Think about it—a culture of people who are excited about evolving the organization to the next level. What does that take? What does that look like? How would those people behave?

To break it down, we know that people react to change a lot better when they know what to expect and why. This suggests transparency. And, we also know that people support change when they help build the solution. This suggests that they have a level of influence in the decision-making process. And finally, we also know that people look for change when they believe it will increase the opportunities for everyone to be successful or advance. This suggests that fair opportunity is present.

A culture of change is one where the attributes include transparency, influence, and opportunity. People respond to these conditions or attributes by working within this culture and taking personal ownership of what they do—resulting in a higher sense of job satisfaction.

What all of this means is that a culture of change is really a partnership between leadership and employees. It means in this partnership, when built in an atmosphere of transparency—where influence and opportunity are readily available—people will be mentally and emotionally capable of tackling change. They will have the capacity for it.

In order to fully understand this culture, let's drill down into each of the attributes further.

Transparency

Transparency is a condition that exists when all parties involved have a clear understanding about the motives, the rationale, the obligations, and the intentions of all other parties. It is a condition of open, honest communication where people share information so as to jointly, collectively, and individually make better decisions—especially when those decisions will have a direct impact on other people.

We're in Trouble

A few years ago, a company they had not been meeting their numbers for some time and was losing money. The management of the company talked about this with the employees and informed them that if the numbers could not be brought up later in the year, some tough decisions would have to be made.

This story illustrates how management was clear and transparent with the employees about the financial health of the company. This is in complete contrast to the company in the following story.

Sorry, We're Closed

A transportation company was struggling financially, but no one knew. Not the customers, not management, not the general public. Only the owner knew.

One day, however, everyone found out in a very big way when people showed up for work and found the gates locked with a big sign on it. That was it. No warning. No notice. No transparency.

Transparency from leadership represents trust. It tells employees that leadership believes that they have the skills and intellect to make tough decisions; and to be privy to information about the organization that helps them to not only do their job, but to do a better job because they can see the big picture.

Opportunity

Fair opportunity is when people do not feel as though they have to climb all over each other to get some recognition from leadership. It is the chance to participate and to contribute. It is the idea that advancement is within the control of the person and not someone or something else.

To be fair, advancement is not necessarily the same for everyone. It is more than just getting a promotion to the next level. It is about getting the chance to explore innate personal talents and skills and then to contribute those to the betterment of the organization—because, to those within the organization, it is a community.

When fair opportunity exists, people support one another and conflict is low. Everyone wins.

Influence

Influence is when people feel as though they have some level of control in what happens to them and around them. Power, on the other hand, is when people feel as though they have control over others. And collective or shared influence is when people feel a sense of shared responsibility for the community they are all a part of.

Collective influence is critical in creating a culture of change. It is not enough for people to want to control what happens to themselves, they have to also feel a sense of shared responsibility for the well-being of those in the organization. This is where true collaboration is born.

Let's Fix This Together

A few years ago, a company was facing a tough decision. It had not been very profitable and they needed to cut costs in order to remain open.

Faced with this decision, the owner did not know what was best and agonized over the idea that they might need to reduce staff in order to keep the business afloat. He took this dilemma to his staff during an all-hands, company-wide meeting.

Those people got involved in making a tough decision. They all agreed to accept a pay cut so that no one would be fired.

In this specific story, the company went on to rebound from its losses and come back stronger than ever. People shared the responsibility for quality, service, and each other. The customers and the community at large responded by supporting the business more.

Personal Ownership

Personal ownership is when people feel as though they are working for their own business. This is not the same thing as people trying to take over or run the organization while pushing others to the side.

Personal ownership happens when they believe strongly in what the organization does and what it stands for; and they feel valued as a key part of its operations. It happens when people feel that they have influence, and they work in a transparent environment with lots of fair opportunity.

People who have a sense of personal ownership hold themselves accountable and take responsibility for change. They jump right in with both feet.

Job Satisfaction

Job satisfaction is when people feel pride in their performance and conduct in the workplace. It's not ego—it's a sense of accomplishment that comes from feeling confident about how well they perform and how their deliverables are received by others.

People who have pride in their work mentor others. They help them adapt to change and they support change when they believe it is the right thing to do for the organization.

LAYERS OF CHANGE

Organizations can adopt a culture of change by incorporating two change layers within their operational framework: strategic and tactical (shown in Figure 12.1). It is important to understand how each of these layers is unique from the other and how they each complement one another.

Strategic Change

Strategic change is the operational function of supporting the business through preparedness and coordination of change management activities across multiple projects. In many ways, it is similar to the project management office or the change advisory board. They each prepare projects for delivery by providing the tools, methods, and support that are most appropriate for use within the organization.

There is a level of governance that goes beyond individual projects. Change must begin during the discussions that something major is about to happen to the organization. This can be a merger or an acquisition, a complete restructuring of the organization, or the development of a brand new product line.

The purpose behind strategic change is to ensure that tactical changes are successfully adopted and the organization is actually transformed as a result. This begins with determining how to prepare people for change, how to announce it, how to manage it, and especially, how to learn from it.

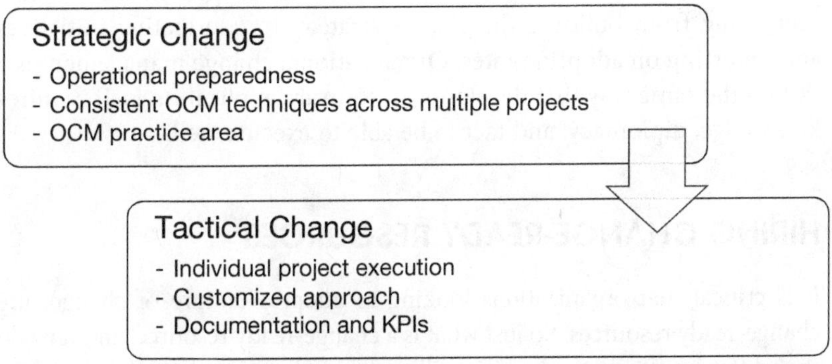

> **Strategic Change**
> - Operational preparedness
> - Consistent OCM techniques across multiple projects
> - OCM practice area

> **Tactical Change**
> - Individual project execution
> - Customized approach
> - Documentation and KPIs

Figure 12.1 Organizational layers of change

The best way to establish strategic change is to build a strategic change group within the organization. This can be formal or informal in the same way that a network or a community of practice is set up.

This group should be tasked with overseeing the delivery and execution of change management within the organization and reporting on its activities to the executive. This includes the establishment of internal best practices and a guiding methodology to ensure consistent delivery. They will be involved at the outset of every major transformation to support the planning of the individual initiatives required to make that vision a reality.

Tactical Change

Tactical change, on the other hand, is the incremental project-based change that occurs on individual projects. It takes guidance from and reports up to the strategic change level.

Tactical change is important because it enables the customization of best practices and the guiding methodology for specific projects in order to achieve particular results. The real question is: does change management need to get involved in every single project that an organization undertakes? The short answer is no.

It is important to deploy tactical change onto projects that will change more than 10% of a particular function, those that result in job losses (even if it is just one), and those that change systems. Does that mean it needs to be a very large endeavor? No, it simply means that change management should be involved to some degree to ensure that people are comfortable and confident enough to move ahead and adopt the changes.

How Is It Executed?

Tactical change is executed by applying the techniques in this book—everything from building the change strategy to conducting initiatives and reporting on adoption rates. Organizational change management is a skill in the same way that developing a software application is. It requires knowledge, diplomacy, and tact to be able to execute well.

HIRING CHANGE-READY RESOURCES

It is critical that organizations looking to adopt a culture of change hire change-ready resources. So just what is a change-ready resource, and how do they differ from anyone else who will undoubtedly be applying for the job?

Characteristics of Change-ready Resources

There are specific characteristics that determine if people are change-ready. These characteristics are unique to the skills required to do the job or to perform the role with any level of proficiency. It is important to look for and to hire new resources for both when attempting to adopt a culture of change as this will help to avoid some of the change fatigue. These characteristics include:

- Open communication
- Active participant
- Open-minded
- Capable of keeping their eye on the results, not the deliverables
- Understands the difference between a dynamic process and an inconsistent process that lacks governance
- Demonstrated track record of change stamina

Olympic Track Team

Imagine that you are the coach of the Olympic track team. You must select athletes who have the skill to compete against the best from every other country that will be represented. Suddenly, you find that you are short a couple of long distance runners. Would you ask two of the relay runners or sprinters to fill in, or would you go on a search to locate people who are not only skilled athletes, but adept at distance running?

The answer is obvious. Of course, you would go on a search to locate and fill the openings with qualified long-distance runners because they have the stamina for distance running that many sprinters and relay runners simply do not have.

In the example, it is easy to see that stamina makes all the difference. In business, however, organizations often make-do with those people that they have on the bench or the people who apply. While this is not always a mistake, it is important to validate that people have the skill and the temperament to adapt to a dynamic working environment so that they do not get change fatigue and quit within the first six months.

Another common mistake that is made in hiring for a dynamic organization is that those in charge most often screen applicants for the role and the company as it is today, and not the transforming or changed organization. In other words, they hire for the organization that they *have* and not the organization that they *want*.

CHANGE FATIGUE

Change fatigue is the mental state of exhaustion that people experience from dealing with overwhelming amounts of change that appear to have no rhyme or reason. Most people describe change fatigue as too much never-ending change.

Why Is Change Fatigue Important to Consider?

A lack of change fatigue is one of the crucial success factors in both implementing new changes and in gaining buy-in from those who are impacted by those changes. When people have zero energy and they do not see the results of the changes that were just implemented, they begin to tune out any new ideas coming down the pipes.

Why Bother?

A large insurance company was known for implementing new incentive programs every couple of years. Unfortunately, they did this because the programs were never fully implemented, usually ineffective, and people did not buy in to them.

The problem was that people who initially bought in to the first few programs were disappointed by how difficult it was to earn incentives, and as soon as they had figured that out, the program changed.

Can Change Fatigue Be Avoided?

Change is an inevitable part of life—change fatigue is not. The easiest way to avoid change fatigue is to masquerade it as something else so people do not see it. But, of course, that is not always easy and not always practical.

That being said, there are two other ways in which to avoid change fatigue in a dynamic organization: get everyone personally invested in the changes, and then conduct change in waves. Throughout this book, there have been descriptions about how to get people personally invested in change. It cannot be stressed enough. Getting every single person who is impacted by change invested in the outcomes is *the* objective.

Conducting change in waves is the critical difference between a culture of change and a culture that changes. In doing so, not everyone is involved in the changes at the same time. In that way, people actually get breaks from the process of change. It is important to remember that *everyone* needs some sense of stability and a feeling of control. It is in between the waves of transformation that they get these.

INDEX

Note: Page numbers followed by "*f*" and "*t*" indicate figure and table respectively.